Flow Simulation Using SOLIDWORKS 2023

CADCIM Technologies
525 St. Andrews Drive
Schererville, IN 46375, USA
(www.cadcim.com)

Contributing Author
Sham Tickoo
Professor
Department of Mechanical Engineering Technology
Purdue University Northwest
Hammond, Indiana, USA

CADCIM Technologies

Flow Simulation Using SOLIDWORKS 2023
Sham Tickoo

CADCIM Technologies
525 St Andrews Drive
Schererville, Indiana 46375, USA
www.cadcim.com

ISBN 978-1-64057-220-1

www.cadcim.com

DEDICATION

*To teachers, who make it possible to disseminate knowledge
to enlighten the young and curious minds
of our future generations*

*To students, who are dedicated to learning new technologies
and making the world a better place to live in*

THANKS

*To the faculty and students of the MET Department of
Purdue University Northwest for their cooperation*

To employees of CADCIM Technologies for their valuable help

Online Training Program Offered by CADCIM Technologies

CADCIM Technologies provides effective and affordable virtual online training on various software packages including Computer Aided Design, Manufacturing and Engineering (CAD/CAM/CAE), computer programming languages, animation, architecture, and GIS. The training is delivered 'live' via Internet at any time, any place, and at any pace to individuals as well as the students of colleges, universities, and CAD/CAM/CAE training centers. The main features of this program are:

Training for Students and Companies in a Classroom Setting

Highly experienced instructors and qualified engineers at CADCIM Technologies conduct the classes under the guidance of Prof. Sham Tickoo of Purdue University Northwest, USA. This team has authored several textbooks that are rated "one of the best" in their categories and are used in various colleges, universities, and training centers in North America, Europe, and in other parts of the world.

Training for Individuals

CADCIM Technologies with its cost effective and time saving initiative strives to deliver the training in the comfort of your home or work place, thereby relieving you from the hassles of traveling to training centers.

Training Offered on Software Packages

CADCIM provides basic and advanced training on the following software packages:

CAD/CAM/CAE: CATIA, Pro/ENGINEER Wildfire, Creo Parametric, Creo Direct, SOLIDWORKS, Autodesk Inventor, Solid Edge, NX, AutoCAD, AutoCAD LT, AutoCAD Plant 3D, Customizing AutoCAD, EdgeCAM, and ANSYS

Architecture and GIS: Autodesk Revit (Architecture, Structure, MEP), AutoCAD Civil 3D, AutoCAD Map 3D, Primavera, and Bentley STAAD Pro

Animation and Styling: Autodesk 3ds Max, Autodesk Maya, Autodesk Alias, The Foundry NukeX, and MAXON CINEMA 4D

Computer Programming: C++, VB.NET, Oracle, AJAX, and Java

*For more information, please visit the following link: **https://www.cadcim.com***

Note

If you are a faculty member, you can register by clicking on the following link to access the teaching resources: ***https://www.cadcim.com/Registration.aspx***. The student resources are available at ***https://www.cadcim.com***. We also provide **Live Virtual Online Training** on various software packages. For more information, write us at ***sales@cadcim.com***.

Table of Contents

Chapter 2: Introduction to SOLIDWORKS Flow Simulation

Chapter 3: Creating and Preparing Model for Flow Simulation

Chapter 4: Creating a Flow Simulation Project

Chapter 5: Checking Geometry

Chapter 6: Boundary Conditions

Chapter 7: Creating Goals

Chapter 8: Analyzing Results

This page is intentionally left blank

Preface

Flow Simulation Using SOLIDWORKS 2023

SOLIDWORKS Flow Simulation, originally developed by SOLIDWORKS Corporation and subsequently acquired by Dassault Systèmes, is a powerful computational fluid dynamics (CFD) software solution. It offers a comprehensive suite of tools for simulating and analyzing fluid flow, heat transfer, and other related phenomena in engineering and design applications.

This CFD software is renowned for its robust capabilities in simulating fluid behavior within 3D models, enabling engineers and designers to gain valuable insights into how fluids interact with their designs. SOLIDWORKS Flow Simulation is a parametric, feature-based tool that seamlessly integrates with SOLIDWORKS 3D modeling, allowing users to explore and optimize designs for various fluid dynamics scenarios.

SOLIDWORKS Flow Simulation is highly user-friendly, making it accessible to both novice and experienced users. It provides a wide range of features and options for setting up and running simulations, including defining boundary conditions, specifying fluid properties, and visualizing results in an intuitive and interactive manner. This software is particularly valuable for engineers working on projects involving airflow, thermal management, HVAC systems, liquid cooling, and a myriad of other applications where fluid behavior plays a crucial role in design performance.

In addition to its simulation capabilities, SOLIDWORKS Flow Simulation facilitates the generation of comprehensive reports and visualizations, allowing users to communicate their findings effectively to stakeholders and make informed design decisions. Whether you are optimizing product performance, evaluating thermal management strategies, or analyzing the flow characteristics of your designs, SOLIDWORKS Flow Simulation is a versatile and indispensable tool for engineering and design professionals.

Flow Simulation Using SOLIDWORKS 2023 textbook has been meticulously crafted to assist individuals eager to delve into the world of fluid dynamics and computational fluid dynamics (CFD) using SOLIDWORKS Flow Simulation. This comprehensive guide leverages real-world mechanical engineering industry examples and step-by-step tutorials to bridge the gap between theoretical knowledge and practical application.

Some of the main features of the textbook are as follows:

- **Tutorial Approach**
  The author has adopted the tutorial point-of-view and the learn-by-doing approach throughout the textbook. This approach guides the users through the process of creating the models in the tutorials.

- **Real-world Mechanical Engineering Projects as Tutorials**
  The author has used the real-world mechanical engineering projects as tutorials in this textbook so that the readers can correlate the tutorials with the real-time models in the mechanical engineering industry.

- **Notes**
  Additional information related to various topics is provided to the users in the form of notes.

- **Learning Objectives**
  The first page of every chapter summarizes the topics that are covered in the chapter.

- **Self-Evaluation Test, Review Questions, and Exercises**
  Each chapter ends with Self-Evaluation Test that enables the users to assess their knowledge of the chapter. The answers to Self-Evaluation Test are given at the end of the chapter. Also, the Review Questions and Exercises are given at the end of the chapters and they can be used by the Instructors as test questions and exercises.

- **Heavily Illustrated Text**
  The text in this textbook is heavily illustrated with the help of line diagrams and screen captures.

Symbol Used in this Textbook

Note

The author has provided additional information to the users about the topic being discussed in the form of Notes.

Formatting Conventions Used in the Textbook

Please refer to the following list for the formatting conventions used in this textbook.

- Names of tools, buttons, options, toolbars, and are written in boldface.

  Example: The **Extrude Boss/Base** tool, the **Mid-Plane** option, the **OK** button, the **Features** toolbar, and so on.

- Name of CommandManager, PropertyManager, rollouts, dialog box, drop-down lists, spinners, selection boxes, areas, edit boxes, check boxes, and radio buttons are written in boldface.

  Example: The **Features CommandManager**, the **Boss-Extrude PropertyManager**, the **Open** dialog box, the **End Condition** drop-down list, the **Depth** spinner, the **Direction of Extrusion** selection box, the **Draft outward** check box, and so on.

- Values entered in edit boxes are written in boldface.

  Example: Enter **5** in the **Radius** edit box.

- Names and paths of the files are written in italics

  Example: *C:\Documents\SOLIDWORKS\c08\ c08_tut01*

Naming Conventions Used in the Textbook
Tool

If you click on an item in a toolbar and a command is invoked to create/edit an object or perform some action, then that item is termed as tool.

For example:
To Create: **Line** tool, **Smart Dimension** tool, **Extruded Boss/Base** tool
To Modify: **Fillet** tool, **Draft** tool, **Trim Surface** tool
Action: **Zoom to Fit** tool, **Pan** tool, **Copy** tool

If you click on an item in a toolbar and a dialog box is invoked wherein you can set the properties to create/edit an object, then that item is also termed as tool, refer to Figure 1.

For example:
To Create: Extruded Boss/Base tool, **Mirror** tool, **Rib** tool
To Modify: Flex tool, **Deform** tool

In this textbook, the path to invoke a tool is given as:

CommandManager: Features > Extruded Boss/Base
SOLIDWORKS Menus: Insert > Boss/Base > Extrude
Toolbar: Features > Extruded Boss/Base

Flyout

A flyout is the one in which a set of tools are grouped together. You can identify a flyout with a down arrow on it. A flyout is given a name based on the types of tools grouped in it. For example, **Line** flyout, **View Settings** flyout, **Fillet** flyout, and so on; refer to Figure 1.

*Figure 1 The **Line**, **View Settings**, and **Fillet** flyouts*

PropertyManager

The naming conventions for the components in a PropertyManager are mentioned in Figure 2.

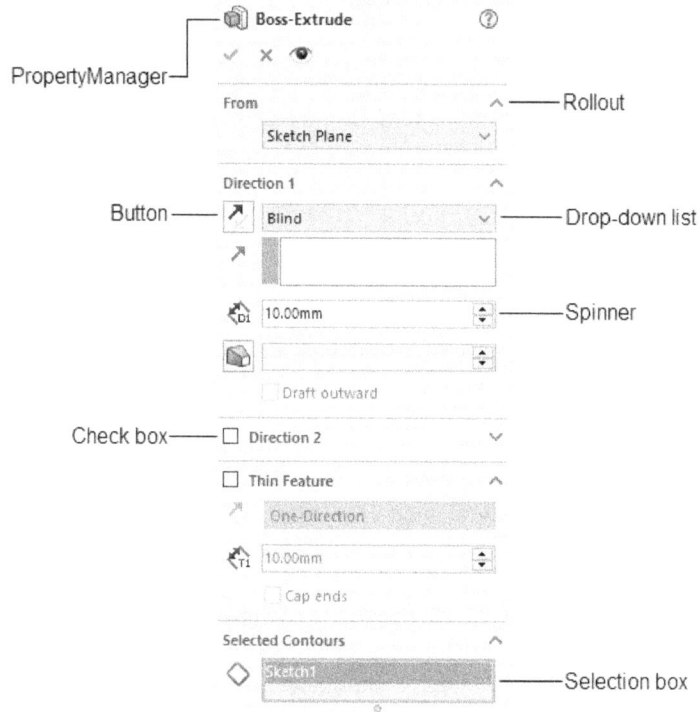

*Figure 2 The **Boss-Extrude PropertyManager***

Button

The items in a dialog box that has a 3D shape like a button is termed as **Button**. For example, **OK** button, **Cancel** button, and so on.

Free Companion Website

It has been our constant endeavor to provide you the best textbooks and services at affordable price. In this endeavor, we have come out with a Free Companion website that will facilitate the process of teaching and learning of Flow Simulation Using SOLIDWORKS 2023. If you purchase this textbook, you will get access to the files on the Companion website.

The following resources are available for the faculty and students in this website:

Faculty Resources
• **Tutorial Files**
  The tutorial files used in examples are available for free download.

• **Input Files**
  The input files used in examples are available for free download.

- **Technical Support**
  You can get online technical support by contacting *techsupport@cadcim.com*.

- **Instructor Guide**
  Solutions to all review questions and exercises in the textbook are provided in the instructor guide to help the faculty members test the skills of the students.

To access the files, you need to register by visiting the **Resources** section at *www.cadcim.com*.

Student Resources
- **Tutorial Files**
  The tutorial files used in examples are available for free download.

- **Input Files**
  The input files used in examples are available for free download.

- **Technical Support**
  You can get online technical support by contacting *techsupport@cadcim.com*.

If you face any problem in accessing these files, please contact the publisher at *sales@cadcim.com* or the author at *Stickoo@pnw.edu* or *tickoo525@gmail.com*.

Video Courses
CADCIM offers video courses in CAD, CAE Simulation, BIM, Civil/GIS, and Animation domains on various e-Learning/Video platforms. To enroll for the video courses, please visit the CADCIM website using the link *https://www.cadcim.com/video-courses*.

Stay Connected
You can now stay connected with us through Facebook and Twitter to get the latest information about our textbooks, videos, and teaching/learning resources. To stay informed of such updates, follow us on Facebook *(www.facebook.com/cadcim)* and Twitter (*@cadcimtech*). You can also subscribe to our YouTube channel *(www.youtube.com/cadcimtech)* to get the information about our latest video tutorials.

This page is intentionally left blank

Chapter 1

Introduction to Computational Fluid Dynamics(CFD)

Learning Objectives

After completing this chapter, you will be able to:

- *Get familiar with the basic concepts of CFD*
- *Understand the different phases of CFD simulation*
- *Know about the concept of flow*
- *Visualize the flow*
- *Understand the governing equations of fluid flow*
- *Get familiar with the concept of heat transfer*
- *Know about the applications of CFD*

INTRODUCTION

Simulation is used by the engineers to predict the behavior of a product under certain boundary conditions. If you are not able to predict the behavior of a new product accurately, then it can have an adverse effect on the company's revenue. To minimize the risk of product failure, you should know how to predict the behavior of the product during working condition. Computational Fluid Dynamics(CFD) is one of the simulation techniques that can help you to predict the behavior of a component under given conditions.

Computational fluid dynamics was developed to solve the problems regarding the heat and mass transfer, aeronautics, vehicle aerodynamics, chemical engineering, industrial design, nuclear design, and many other domains of engineering. In this methodology, the problems of a complex geometry and boundary conditions are solved on the basis of fundamental equations of a fluid motion. Apart from it, you need numerical solution technique to apply these equations to solve the problems in digital manner. During 1960s, scientists discovered a way to solve the fluid motion equations in digital computers. The scientist realized that the CFD tool would be helpful for them in the future in solving the fluid dynamics problems with low cost and greater flexibility solution.

Computational fluid dynamics (CFD) refers to simulating, analyzing, and solving fluid flow problems using numerical algorithms and computer-based analysis. Its evolution parallels that of fast, low-cost computers, which are now crucial for a wide range of human needs. However, the necessity for even faster and high-performing computers at the same or lower cost, in order to conduct big and complex 3D simulations on an average home computer, is less widely recognized. The availability of low-cost high-performing computing technology and the development of user-friendly interfaces have encouraged new interest in CFD, which has been used in industry since the 1990.

CFD has increasingly acted as an alternative and/or supplementary tool to more traditional methods in the systematic investigation of various controls in river morphology, flow structure, and sediment transport, and it has played an important role in river management and flood prediction. CFD simulations may provide more insight into, and a clearer explanation of, the structure of the flow and the interactions of the fundamental components of the processes than traditional field and/or laboratory data.

The use of CFD in the design of industrial products and processes is becoming increasingly common. CFD offers a number of distinct advantages over experiment-based approaches to fluid system design:

• CFD simulations are affordable, and as computers become more powerful, costs are projected to drop.

• CFD simulations can be completed in a relatively short time.

• The designer can analyze any point in the region of interest using CFD and assess its performance using a set of thermal and flow metrics.

• CFD allows you to simulate any physical situation hypothetically.

Engineering simulation technique advancements have resulted in significant improvements in product design process. It is now possible to develop a virtual product in a computer, study its performance, recommend design changes, and check performance or cost savings. It gives an ideal platform for new product development breakthroughs. For established products, however, it leads to significant increases in performance and efficiency. Computer simulation and analysis have been an important aspect of engineering studies in recent years.

PHASES OF CFD SIMULATION

CFD simulation has three main phases named as pre-processing, solver, and post-processing. The phases are discussed next.

Pre-processing

Pre-processing is the initial step in CFD simulation, and if done correctly, it can help in accurately defining the simulation's parameters. In this phase, the following things are taken into consideration for performing CFD simulation:

- The region of geometry where you want to perform simulation

- Meshing the geometry

- Creating and assigning materials

- Selection of the physical and chemical processes

- Defining fluid properties

- Specifying suitable boundary conditions

Similar to meshing in finite element modeling, meshing CFD adds a numerical mesh to the fluid body and boundary. The precision of the CFD simulation is determined by mesh algorithms, which are used to construct the sets of mesh nodes. By altering mesh density in CFD modeling, it is possible to strike a compromise between solution accuracy and calculation time. Analysts must be very attentive about cell type, amount of cells, and calculation time in order to make appropriate conclusions and produce dependable results. To improve the accuracy of CFD analysis, an ideal grid design is required.

Solver

The basic objective of computational fluid dynamics (CFD) is to solve the governing equations for flow physics problems. Once the problem physics has been identified, the fluid material characteristics, flow physics model, and boundary conditions are developed that need to be solved. This can be accomplished by using readily available commercial software, each of which has its unique set of features. The computational approach takes into consideration the following points:

a) Integration of the fluid flow controlling equations over all of the domain's control volumes.

b) Converting the resulting integral equations into a system of algebraic equations.

c) Iterative solution of the algebraic equations.

Post-processing

Post-processor is the next step to analyze the outcomes obtained during the simulation stage. In this stage, you use accessible tools such as vector plots, contour plots, data curves, and streamlines. You will obtain accurate reports and graphical representations this way.

FLOW

The fluid and flow have different characteristics. Viscosity, density, surface tension, diffusivity, and heat conduction are inherent characteristics of fluids that may be characterized as functions of temperature, pressure, and composition. Pressure, turbulence, and turbulent viscosity are properties that are affected by the flow. The flow is classified into the following categories:

a) Laminar Flow and Turbulent Flow

b) Single Phase Flow and Multiphase Flow

c) Steady or unsteady

d) Compressible or incompressible

e) Viscous or non-viscous

f) Rotational or irrotational

Laminar Flow

Laminar flow is defined as the flow in which fluid particles move in a smooth or regular pattern. As a result, laminar flow is also known as streamline flow or viscous flow. In laminar flow, the velocity of fluid particles remains constant at all points within the fluid. Laminar flow is a simplified and unrealistic model of fluid flow; however, it is often used as an initial approximation of real fluid flow.

The Reynolds number (Re) is a dimensionless quantity that is used to characterize the flow of fluids. It was introduced in 1883 by Osborne Reynolds. The Reynolds number is used to predict the transition from laminar flow to turbulent flow. In laminar flow, the fluid particles move in straight lines and at constant velocity. However, when the Reynolds number exceeds a critical value, the flow becomes turbulent. The Reynolds number in a pipe is defined as:

$$Re \; = \; \frac{\rho V d}{\mu}$$

Where,
Re = Reynolds number
ρ = Density of the fluid
V = Flow speed
d = Internal diameter of pipe
μ = Dynamic viscosity of the fluid

The Navier-Stokes equations explain the momentum transmission of a flow dominated by viscous forces in laminar flow. For single-phase systems, CFD can produce extremely accurate flow models as long as the flow remains laminar.

Turbulent Flow

Turbulent flow is defined as a flow with significantly irregular streamlines. Turbulent flow is caused by a high degree of mixing between the fluid and its surroundings, resulting in a chaotic flow pattern. Turbulent flow is extremely difficult to forecast due to its unpredictability. Many natural phenomena, such as wind and water currents, show turbulent movement.

The Reynolds number is a dimensionless parameter that describes how important inertial forces are in comparison to viscous forces. When water flows through a pipe, the pipe is generally considerably bigger than the water molecules, thus inertial forces are insignificant and viscous forces dominate the flow. As a result, the Reynolds number is low, and the flow is described as laminar. In this case, it can be shown that viscosity is proportional to the reciprocal of the Reynolds number. Thus, if the Reynolds number increases, viscosity decreases.

Turbulent flow occurs when the forces acting on the fluid are large. Turbulent flow is characterized by chaotic and uneven fluid motion. A tornado's air movement is an example of turbulent flow. In this flow, the Reynolds number is really high and the fluid travels extremely quickly. The air velocity of a tornado may reach up to 161 kilometers per hour.

The transition from laminar to turbulent flow occurs at the critical Reynolds number. This is the point where a fluid changes from laminar to turbulent flow. The flow is laminar when the Reynolds number is smaller than 2000. Flow becomes unstable when Reynold number lies between 2000 and 3000, means it can be laminar but can also be turbulent due to tiny obstacles and surface roughness, and it can cycle arbitrarily between laminar and turbulent states. In fact, chaotic behavior may be seen in the flow of a fluid that has a Reynolds number between 2000 and 3000. When a system's behavior is so sensitive to one aspect that it is impossible to anticipate, it is said to be chaotic. When a fluid's Reynold's number falls in this range, it is difficult, but not impossible, to forecast whether the flow will be turbulent or not because of the very sensitive dependency on parameters like roughness and obstacles on the character of the flow. The flow is nonlinear when a little adjustment in one parameter is made. Flow is turbulent when the Reynold number exceeds 3000.

Single-phase Flow

You can get highly precise solutions in single-phase laminar flow, as well as adequate flow simulations in turbulent flow in most instances. The division of the flow into a number of distinct pieces, or cells, and then using an average velocity for each cell is a typical technique to turbulence modeling. By dividing the amount of fluid in the cell by the area of the cell, the average velocity can be computed. Within each cell, the average velocity is frequently considered to be constant, which substantially simplifies the mathematics required in turbulence modeling. The number of cells in the grid, as well as the size of each cell, must be specified in order to apply this strategy. The number of cells in a uniform grid is the same in every direction. A non-uniform grid, on the other hand, is required in many applications. In a wind tunnel, for example, the grid is often finer in the direction of flow than it is perpendicular to the flow for modeling turbulence. It is therefore necessary to utilise a non-uniform grid.

The main issue typically involves simulating reactant mixing in laminar or turbulent flow for rapid reactions. When the reaction rate is quick relative to mixing, there will be significant concentration gradients that cannot be resolved in the grid, necessitating the use of a mixing model in conjunction with a chemical reaction. Chemical reaction modeling in CFD is a difficult task. It necessitates the integration of a chemical reaction model with a numerical fluid flow simulation. This category includes both gaseous combustion and liquid-liquid ion-ion processes. The Lattice Boltzmann Method is used to create CFD models for chemical reactions (LBM). The LBM is a kinetic approach for simulating large-scale systems. This approach has been used to model combustion systems with great success. This technique, however, assumes that the flow field is unaffected by chemical composition. As a result, it can not predict specific movement in complex settings like flames.

Multiphase Flow

Multiphase flow is a type of complex flow that contains several phases. Chemical engineering, fluid mechanics, and thermodynamics are just a few of the things involved in multiphase flow. Multiphase flow can take many different forms:

a) Flow of gas and liquid (in which the phases are two gases or a gas and a liquid)

b) Flow of liquids into liquids (in which the phases are two liquids)

c) Gas-solid flow (in which the phases are a gas and solid particles)

d) Liquid-solid flow (in which the phases are liquid and solid particles)

e) Porous material with multiphase flow (in which the phases are gas, liquid, and solid)

For a multiphase system comprising very tiny particles, bubbles, or droplets that closely follow the continuous phase, reasonable simulation results can be obtained. However, for systems containing large particles or bubbles, the simulation results are not satisfactory. Systems in which the dispersed phase has a large effect on the continuous phase are more difficult to simulate accurately, and only crude models are available for multiphase systems with a high load of the dispersed phase.

The most simple approach to multiphase flow is to describe the continuous phase as a single phase system and assume that the dispersed phase behaves like a viscous liquid. The dispersed phase is considered to be incompressible and to have a constant volume fraction. The continuous phase might be immiscible or miscible with the dispersed phase. The dispersed phase has a different density than the continuous phase in the case of immiscibility. The scattered phase has the same density as the continuous phase in the case of miscibility.

The assumption of constant volume fraction is valid for most dispersed systems at low to moderate volume fractions. At higher volume fractions, this assumption breaks down and the density of the dispersed phase is no longer equal to that of the continuous phase. This occurs because at high volume fractions, the particles in the dispersed phase begin to interact with each other and form clusters. These clusters have different densities than those of the individual particles. At that moment, the quality of the simulations is limited not by the computer speed or memory but by the lack of excellent models for multiphase flow. For example, the oil-water system is a mixture of two immiscible fluids with different densities and viscosities. These properties are very difficult to model, and they vary widely depending on the composition of fluids.

However, multiphase flows are very important in engineering since many common processes involve multiphase flow. Many applications, such as boiling, heterogeneous catalysis, and distillation, are interested in mass and heat transfer between the phases. The behavior of a fluid with different phases can be characterized by its composition, which is the ratio of mass fractions of each phase. An important quantity for describing the state of a multiphase flow is the so-called "equilibrium curve", which shows how the composition changes as a function of the flow parameters such as temperature and pressure. The equilibrium curve can be used to predict phenomena such as nucleation and coalescence.

Steady Flow

A steady flow has variables that vary from point to point but do not fluctuate over time, such as velocity, pressure, and cross-section.

Unsteady Flow

The flow is regarded as unsteady when the variables in the fluid fluctuate over time at any location.

Compressible Flow

The fluid has the ability to compress or expand. You should also note that a fluid can either be a liquid or a gas. A compressible fluid is the one in which compression and expansion have a substantial impact on its density (kg/m3).

Incompressible Flow

Incompressible fluids are defined as those whose density remains unaffected by compression and expansion. An incompressible fluid's volume remains constant, and its density is assumed to be constant. Consider a liquid contained in a cylinder. The piston will cease moving once it comes into touch with the liquid if the cylinder is sealed. When the piston is retracted, an empty space above the liquid surface is generated. The quantity of space (volume) occupied by the liquid remains constant (actually the volume does change but the change is very tiny). The fluid density (kg/m3) remains steady since the amount of liquid is nearly unchanged. As the density changes caused by pressure and temperature are negligible, liquids are always considered incompressible fluids.

Viscous and Non-Viscous Flow

Viscosity is a property of real fluids. When a force is applied to a fluid, a force resistive to the applied force is also generated. The property of resistivity in fluid flow is known as viscosity, and a fluid with viscosity is referred to as a viscous fluid. Non-viscous fluid, on the other hand, is a virtual fluid that has no viscosity.

Rotational or Irrotational Flow

Irrotational flow is defined as a flow in which from one instant to the next, no element of the flowing fluid experiences any net rotation with respect to a set of coordinate axes. Rotational flow is defined as a flow in which fluid particles spin about their own axis while flowing along streamlines.

FLUIDS

The term fluid refers to a material that has no defined structure and quickly yields to external pressure. For example, a gas is a fluid. It has no definite shape and can be compressed easily by external pressure. A liquid is also fluid but has a definite shape, although it can be compressed.

Fluids can also be classified into following categories:

a) Ideal Fluids

b) Real Fluids

c) Newtonian Fluids

d) Non-Newtonian Fluids

e) Ideal Plastic Fluids

These fluids are discussed next.

Ideal Fluids

Ideal Fluids are not compressible and have no viscosity, which allows them to flow freely without resistance. This imaginary fluid category is employed in various computations. In computational fluid dynamics, ideal fluids are used as an approximation for real fluids. For example, the Navier Stokes equations governing the motion of a viscous fluid may be simplified by approximating the viscosity as zero.

Real Fluids

Real Fluids are fluids that exist in the real world. The viscosity of these fluids is determined by their composition. In certain circumstances, it is compressible by nature. Some of the common examples of real fluids are water, blood, honey, syrup, and molasses.

Newtonian Fluids

Newtonian fluids have a shear stress that is directly proportional to the shear strain. Water, gasoline, alcohol, and other Newtonian fluids can be Newtonian fluids in a certain temperature range. The viscosity of a Newtonian fluid is independent of the shear direction.

Non-Newtonian Fluids

In Non-newtonian fluids, the shear stress is not directly proportional to the shear strain. For example, in cornstarch and water, which are a mixture of a Newtonian fluid (water) and a

Non-newtonian fluid (cornstarch), when you increase the rate of shear, the viscosity decreases. The viscosity of Non-newtonian fluids depends on the shear direction.

Ideal Plastic Fluids

Ideal plastic fluids have the same properties as those of an ideal fluid except for viscosity. This makes them much easier to handle than non-ideal fluids. Ideal plastic fluids are used in hydraulic circuits and hydraulic machinery because they are able to withstand large shear forces without failing. They also have high flow rates, making them suitable for a wide range of applications.

VISUALIZING FLOW

When you try to visualize a flowing fluid pattern, your imagination may fail. Even with the most powerful computers now available, you cannot find out all details of flows from the governing equations. Flow is more precisely understandable when shown using streamlines, streaklines, or pathlines.

Streamlines

A streamline may be defined as an imaginary curve or line in the flow field that indicates the direction of the instantaneous velocity at any location. The pattern of streamlines varies from instant to instant in an unsteady flow when the velocity vector changes with time. The pattern and direction of streamlines remains fixed in a steady flow.

Streaklines

The connected line of particles put one after another into a flow is known as a streakline. A fixed point in the flow field defines a streak line. It is particularly interesting in the context of experimental flow visualization. The smoke trail from a chimney is a good example of a streakline.

Pathline

A path line is the path that a single fluid particle follows as it travels over time. As a result, a path line shows the direction of velocity of the same fluid particle at different points in time. The path made by a balloon floating in the air is an example of a pathline.

GOVERNING EQUATIONS

Various characteristics influence how fluid moves in the physical domain. Those characteristics must be clearly described in order to enable a transition between the physical and numerical domains for the aim of bringing the behavior of fluid flow to light and building a mathematical model. The main parameters that should be evaluated while conducting a fluid flow investigation are velocity, pressure, temperature, density, and viscosity. Those characteristics vary greatly depending on physical phenomena such as combustion, multiphase flow, turbulence, mass transport, and so on. These categories can be classified as kinematic, transport, thermodynamic, and other miscellaneous properties.

Computational Fluid Dynamics is governed by the Navier-Stokes equations. It is based on the conservation law of physical properties of fluid. It is also known as the momentum equation and is a vector equation generated by applying Newton's Law of Motion to a flowing fluid. It is also known as the momentum equation and is a vector equation generated by applying Newton's Law of Motion to a flowing fluid.

The fundamental governing equations of fluid dynamics, the continuity, momentum, and energy equation are at the heart of computational fluid dynamics. They are mathematical expressions of three fundamental physical principles that govern all fluid dynamics:

a) Conservation of Mass (Continuity Equation)
b) Conservation of Momentum (Newton's Second Law)
c) Conservation of Energy

Conservation of Mass

One of the three basic fluid fundamental laws is the conservation of mass. The conservation of mass equation for a flowing fluid is based on the principle that the mass of a specific collection of neighboring fluid particles is constant. The volume of fluid particles in a given flow that occupies a specific region is called the material volume V(t). A material volume and its surface move with the local fluid velocity(u) so that no particle enters or leaves it. The conservation of mass for a material volume in a flowing fluid can be written as:

$$\frac{d}{dt}\int_{v(t)} p\left(x,t\right)dv = 0$$

Here, p is the fluid density. The Reynolds transport theorem may be used to visualize the consequences of the above equation for the fluid velocity field. The Reynolds transport theorem is a mathematical treatment of the conditions at a stagnation point. The stagnation point is an interface between two flow regions, where there is no relative motion between the two fluid phases. The Reynolds transport theorem describes the conservation of mass, momentum, and energy in the flow over a wide range of conditions.

$$\frac{dN}{dt}\Big|_{sys} = \frac{\partial N}{\partial t}\Big|_{cv} + \int_{cs} pn(\overrightarrow{V}_r \times \hat{\eta})dA$$

Here, N is the extensive property, n is the N per unit mass, cv is the control volume, dA is the elementary surface area, $\hat{\eta}$ is the unit vector, $\overrightarrow{v_r}$ is the velocity of fluid relative to the control volume across that area.

In the case of conservation of mass, N = m and n becomes1. The equation becomes as follows.

$$\frac{dm}{dt}\Big|_{sys} = \frac{\partial m}{\partial t}\Big|_{cv} + \int_{cs} p(\overrightarrow{V}_r \times \hat{\eta})dA$$

For non-deformable control volume, the equation becomes,

$$\frac{dm}{dt}\Big|_{sys} = \int_{cv}\frac{\partial}{\partial t}(\rho)dv + \int_{cs}\rho(\overrightarrow{V}_r \times \hat{\eta})dA$$

For stationary control volume and a system of fixed mass, the equation becomes

$$\int [\frac{\partial\rho}{\partial t} + \overline{V}.(\rho\overrightarrow{V})]dV = 0$$

and then,

$$\frac{\partial\rho}{\partial t} + \overline{V}.(\rho\overrightarrow{V}) = 0$$

where ρ is density, V is velocity, and $\bar{v}$ is gradient operator. The gradient operator can be written as:

$$\bar{\nabla} = \bar{i}\frac{\partial}{\partial x} + \bar{j}\frac{\partial}{\partial y} + \bar{k}\frac{\partial}{\partial z}$$

The fluid is assumed to be incompressible while the density remains constant, and then continuity is simplified as follows, indicating a steady-state process:

$$\frac{\partial \rho}{\partial t} = 0$$

$$\nabla \cdot \vec{V} = \frac{\partial u}{\partial x} + \frac{\partial v}{\partial y} + \frac{\partial w}{\partial z} = 0$$

Conservation of Momentum

Conservation of momentum states that the momentum of the fluid particles does not change unless a force is applied on it. By using the reynolds transport theorem and newtons second law, the equation becomes as follows.

$$\sum \vec{F}_{res} = \frac{d}{dt}(m\vec{V})_{sur} = \int_{cv}\frac{\partial}{\partial t}(\rho\vec{V})dV + \int(\rho\vec{V})(\vec{V}\cdot\hat{\eta})dA$$

where F_{res} is the resultant force acting on the system.

The resulting force F_{res} on the control volume is the sum of the surface force, F_{sur} (pressure, viscous stress), and the body force, F_b, (gravity).

$$\vec{F}_{res} = \vec{F}_{sur} + \vec{F}_b$$

The surface force F_{sur}, can be represented as

$$F_{sur} = \int_{cs} \tau \cdot dA$$

where τ is the stress tensor.

The body force Fb can be represented as

$$F_b = \int_v \rho b \, dV$$

where b is body force per unit mass.

Conservation of Energy

The first law of thermodynamics is a statement of conservation of energy principle. This principle states that energy cannot be created or destroyed, but it can change form. The first law states that the total amount of energy in an isolated system remains constant and is equal to the sum of net rate of energy addition by heat transfer and the net rate of energy addition by work done on the system. The equation to represent that are as follows:

$$\frac{DE}{Dt} = \frac{\delta Q}{\delta t} - \frac{\delta W}{\delta t}$$

Here, E stands for total stored energy, which includes kinetic, potential, and internal energy. One form of energy equation that is commonly used is:

$$\rho \left[\frac{\partial h}{\partial t} + \nabla \cdot (hV) \right] = -\frac{\partial p}{\partial t} + \nabla \cdot (k\nabla T) + \phi$$

Here, k is thermal conductivity and h is enthalpy.

Navier-Stokes Equation

The Navier-Stokes equations are a popular mathematical model for analyzing changes in properties during dynamic or thermal interactions. The equations are based on the ideas of mass, momentum, and energy conservation and are adjustable in terms of the problem's topic. The Navier-Stokes equation in its most generic form:

$$\rho \left(\frac{\partial u}{\partial t} + u\frac{\partial u}{\partial x} + v\frac{\partial u}{\partial y} + w\frac{\partial u}{\partial z} \right) = \rho g_x - \frac{\partial P}{\partial x} + \mu \left(\frac{\partial^2 u}{\partial x^2} + \frac{\partial^2 u}{\partial y^2} + \frac{\partial^2 u}{\partial z^2} \right)$$

$$\rho \left(\frac{\partial v}{\partial t} + u\frac{\partial v}{\partial x} + v\frac{\partial v}{\partial y} + w\frac{\partial v}{\partial z} \right) = \rho g_y - \frac{\partial P}{\partial y} + \mu \left(\frac{\partial^2 v}{\partial x^2} + \frac{\partial^2 v}{\partial y^2} + \frac{\partial^2 v}{\partial z^2} \right)$$

$$\rho \left(\frac{\partial w}{\partial t} + u\frac{\partial w}{\partial x} + v\frac{\partial w}{\partial y} + w\frac{\partial w}{\partial z} \right) = \rho g_z - \frac{\partial P}{\partial z} + \mu \left(\frac{\partial^2 w}{\partial x^2} + \frac{\partial^2 w}{\partial y^2} + \frac{\partial^2 w}{\partial z^2} \right)$$

BOUNDARY CONDITIONS

In order to formulate the CFD problem, you have to specify the geometry of interest and appropriate boundary conditions. All input and outlet conditions, as well as conditions on walls and other boundaries, must be specified. No numerical system is complete without appropriate ways for imposing boundary conditions. The majority of boundary conditions may be classified as steady-state or transient. Throughout the simulation, steady-state boundary conditions are maintained. Transient boundary conditions change over time and are frequently used to model events or cyclical phenomena.

HEAT TRANSFER

Heat transfer is the transfer of thermal energy between two bodies occurred due to temperature difference. It can happen through any material, such as solid, liquid, or gas. In other terms, heat transmission happens when two things have different temperatures.

The three primary modes of heat transfer are conduction, convection, and radiation. Physical models that just include conduction or convection are the easiest to understand. They probably are the most simple to solve analytically. Radiative heat transmission is more complicated, and as a result, it is difficult to solve analytically. Many factors that are difficult to regulate or impact heat transfer by radiation in real-world conditions are difficult to forecast. As a result, numerical approaches are frequently used to solve issues involving heat transfer by radiation.

SOLIDWORKS Flow Simulation will determine the form of energy equation based on heat transfer techniques you have defined, depending on the problem you are seeking to address. Based on the thermal characteristics of each material, surface temperature, and ambient temperature, SOLIDWORKS Flow Simulation determines heat transfer to and from the solid model. For each cell in the grid, the energy equation is solved, and you use the results to update the temperature at each node. Convection, conduction, radiation, and combinations of these heat transfer processes may all be calculated using SOLIDWORKS flow simulation. Additional parameters, like heat capacity, density, thermal conductivity, and specific heat, are calculated using the resultant temperatures.

Conduction

Conduction refers to energy transfer across a system boundary because of a temperature difference by the mechanism of intermolecular interactions. Conduction needs matter and does not require any bulk motion of matter. Conduction is a mechanism of heat transfer. It involves the movement of particles, and hence it is also called kinetic energy transfer. Kinetic energy means the energy of motion, which may be potential or kinetic depending on whether the object is at rest or in motion. When heat flows by conduction, particles of matter (molecules) move past each other as they vibrate about their mean positions. If one end of a block of metal is kept hot and the other end cold, heat will flow by conduction from the hot end to the cold end. In this case, there is a temperature difference between the two ends of the block and it is called a heat differential.

The conduction rate equation is described by using the Fourier Law which states that the heat flow rate is proportional to the temperature gradient and inversely proportional to the distance between the surfaces. This is the conduction law for conductors. Conduction rate equation are as follows:

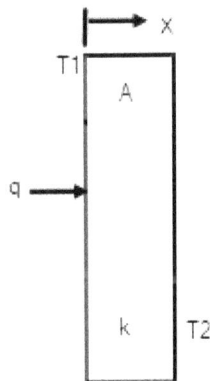

$$q = -kA\overline{\nabla}T$$

In Where,

$$q = \text{Heat Flow vector, W}$$

$$k = \text{Thermal conductivity, W/mK}$$

A = Cross sectional area in direction of heat flow, m²

$$\overline{VT} = \text{Gradient of temperature, K/m} = \frac{\partial T}{\partial x}i + \frac{\partial T}{\partial y}j + \frac{\partial T}{\partial z}k$$

The equation in X-direction,

$$q_x = -kA_x\frac{dT}{dx}$$

The equation in radial direction,

$$q_r = -kA_r\frac{dT}{dr}$$

Convection

Convection is one of the most common ways of heat transfer. Heat is transferred by convection in fluids that are able to take part in bulk flow. Convection can occur when the fluid is either a gas or a liquid; it cannot occur in solids. Convection may be used for a variety of things, such as cooling gadgets and keeping people warm on chilly days. The process also aids in the transportation of heat from the Earth's core to the surface. Convection also allows for the circulation of ocean currents, as well as the rising and sinking of air masses.

There are three primary types of convection: forced, free, and natural. Forced convection occurs when the fluid is being actively moved by some means, such as pumps. Free convection involves the movement of fluids because of density differences within the fluid itself. Natural convection results from buoyancy forces within fluids that are heated from below, and this form of convection usually occurs in liquids, but it can also occur in gases.

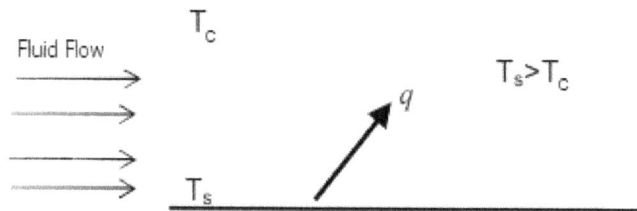

As per the Newton's Law of Cooling, the equation is as follows:

$$q = h\,A_s\,\Delta T$$

Where,

q = heat flow from surface, W

h = heat transfer coefficient, W/m²K

As = Surface area where convection occurs, m²

ΔT = Ts - Tc = Temperature Difference between surface and coolant, K

Typical values of heat transfer coefficient are as follows:

Free convection
gases: 2-28
liquid: 50-105

Forced convection
gases: 28-255
liquid: 50-20,000

Boiling/Condensation: 2400-90,000

Radiation

Radiation is one of the three main ways of heat transfer in matter. The transmission of heat through electromagnetic radiation that arises as a result of the body's temperature is known as radiation heat transfer. Radiation does not necessitate the presence of matter.

According to the Stefan Boltzman law, a specific amount of power (Watts/m²) is emitted from a particular area. This is based on the fourth power of temperature and is referred to as gross heat emission. This fact will be used to write the equation below.

$$E = \sigma \epsilon T_s^4$$

Where,
E = Emissive power of a surface, W/m²

σ = Steffan Boltzman constant, W/m²K⁴

ϵ = Emissivity

T_s = Absolute temperature of the surface, K

The rate of radiation heat exchange between a small surface and a large surrounding is given as follows:

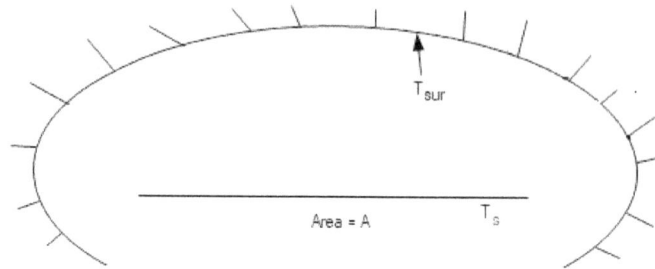

$$q = \sigma \epsilon A(T_s^{\,4} - T_{sur}^{\,4})$$

ϵ = Surface emissivity

A = Surface area

T_S = Absolute temperature of surface, K

T_{Sur} = Absolute temperature of surroundings, K

APPLICATIONS OF CFD

CFD (computational fluid dynamics) is used in both industrial R&D and basic research. It is a tool used in almost every industry. CFD may be used in aerospace engineering, chemical engineering, transportation engineering and biomedical engineering. CFD can be used for both steady state as well as transient problems. Transient analysis can also include unsteady effects such as gusts, turbulence, wind shear, and other unsteady effects. The following is a list of some of the most well-known CFD applications:

- Aircraft design, airfoil and wing design

- Automotive body and chassis design

- Flow boiling in chemical reactors

- Hydrodynamic stability of buildings

- Nonlinear dynamics of electromagnetic systems

- Stresses on off-shore buildings

- Design of ducts and positioning of heating or cooling ducts in a building

Self-Evaluation Test

Answer the following questions and then compare them to those given at the end of this chapter:

1. CFD has _____ phases.

2. A fluid flow is classified as Laminar and _____ flow.

3. The _____ flow contain several phases.

4. The _____ flow has variables that do not fluctuate with time.

5. The _____ fluids are not compressible and have no viscosity.

Review Questions

Answer the following questions:

1. Which of the following is the primary mode of heat transfer?

   (a) Conduction   (b) Convection
   (c) Radiation   (d) All of these

2. Which mode of heat transfer involves Stefan Boltzman law?

   (a) Radiation   (b) Conduction
   (c) Convection   (d) None of these

3. The _____ is the most common way for heat to be transferred in fluids.

4. The _____ visualization is the connected line of particles put one after another into a flow.

5. Newtonian fluid is dependent on shear direction. (T/F)

Answers to Self-Evaluation Test
1. Three, **2.** Turbulent, **3.** Multiphase, **4.** Steady, **5.** Ideal

Chapter 2

Introduction to SOLIDWORKS Flow Simulation

Learning Objectives

After completing this chapter, you will be able to:
- *Understand SOLIDWORKS Flow Simulation*
- *Get familiar with Flow Simulation interface components*
- *Perform meshing in Flow Simulation in SOLIDWORKS*
- *Add tools in Flow Simulation*

INTRODUCTION

SOLIDWORKS Flow Simulation is used to forecast how liquids and gases will behave in given conditions by using computational fluid dynamics simulation tools. This software may be used to evaluate and design fluid flow in a variety of applications. You can also do fluid flow analysis, pressure loss prediction, and turbulence simulation by using this software. It helps you to determine what adjustments need to be made to the design in order to increase the product performance.

Users can create and analyze multiphase flow in a wide range of industries, including energy, transportation, aerospace, marine, and manufacturing. They can also simulate flow using multiple configurations of geometry and control the effects of different variables on the flow field. The user can also save time by importing data from SOLIDWORKS to perform simulations faster. The software supports both parallel computing and GPU for high performance computing.

SOLIDWORKS Flow Simulation reduces the requirement for physical prototypes while providing greater information about the performance of fluid flow models. You can use Flow Simulation to analyze transient or steady-state fluid flow in a model. You can also simulate incompressible or compressible flow, which is the analysis of flow with no change in density or volume. Flow Simulation is based on physics-based simulation models and takes into account any changes in the fluid flow's density, temperature, viscosity, velocity, and pressure.

SOLIDWORKS Flow Simulation stands out for its generally obvious and easy interface, which includes a preprocessor for defining data for the calculation, a co-processor for monitoring and directing the computation, and a postprocessor for displaying the results. As a result, you will have more time to evaluate the data and interact with the calculations.

GETTING STARTED WITH SOLIDWORKS AND SWITCHING TO FLOW SIMULATION

Install SOLIDWORKS 2023 on your system; a shortcut icon of SOLIDWORKS 2023 will automatically be created on the desktop. Double-click on this icon; the system will prepare to start SOLIDWORKS, and after sometime, the **SOLIDWORKS** window will be displayed. On opening SOLIDWORKS for the first time, the **SOLIDWORKS License Agreement** dialog box will be displayed, as shown in Figure 2-1. Choose the **Accept** button in this dialog box; the **SOLIDWORKS 2023** window will open and the **SOLIDWORKS Resources** task pane will be displayed on the right. Also, the **Welcome - SOLIDWORKS** dialog box will be invoked, as shown in Figure 2-2. This window can be used to open a new file or an existing file.

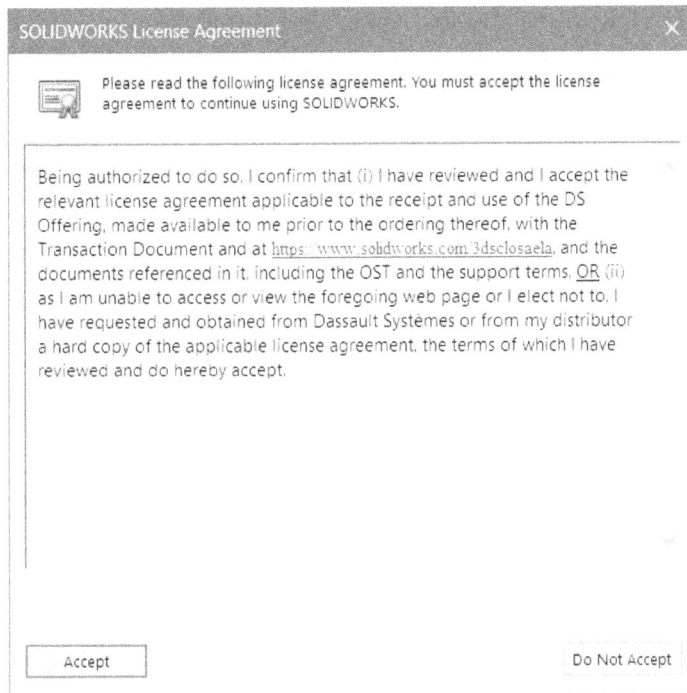

Figure 2-1 The **SOLIDWORKS License Agreement** *dialog box*

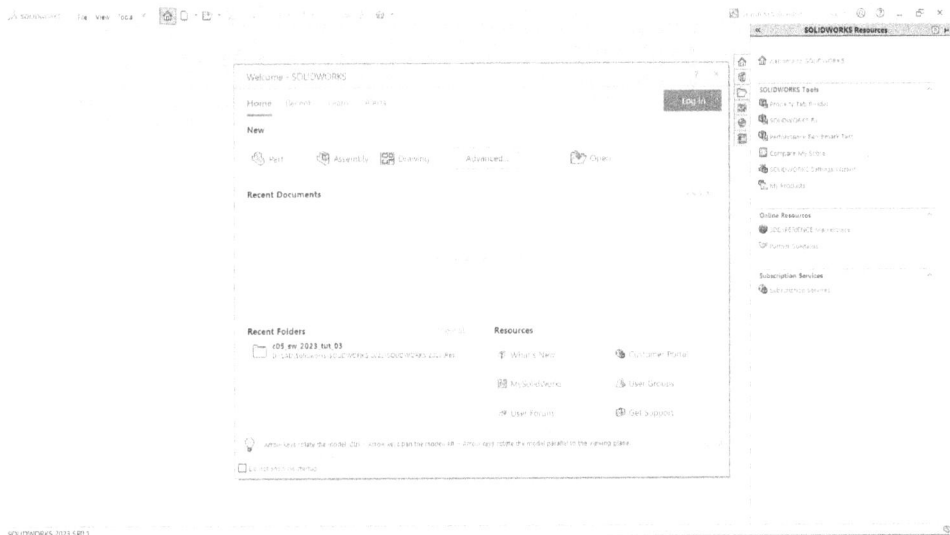

Figure 2-2 The **SOLIDWORKS 2023** *window and the* **SOLIDWORKS Resources** *task pane*

If the **SOLIDWORKS Resources** task pane is not displayed or expanded, choose the **SOLIDWORKS Resources** button located on the right side of the window to display the task pane. This task pane can be used to open online tutorials and to visit the website of SOLIDWORKS partners. Choose the **Part** button from the

Welcome - SOLIDWORKS dialog box or the **New** button from the **Menu Bar** to create a new document. If you start a new document using the **New** button from the **Menu Bar** then the **New SOLIDWORKS Document** dialog box will be displayed, as shown in Figure 2-3.

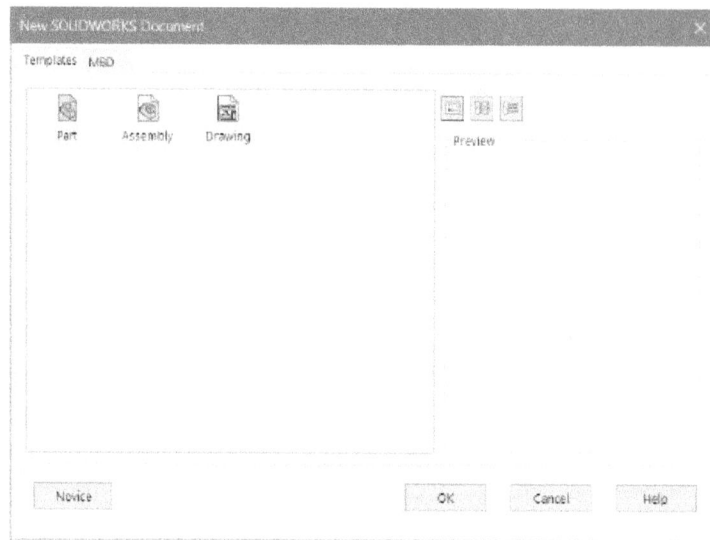

Figure 2-3 The New SOLIDWORKS Document dialog box

Choose the **Part** button to create a part model and then choose the **OK** button from the **New SOLIDWORKS Document** dialog box to enter the Part mode of SOLIDWORKS. Hover the cursor over the **SOLIDWORKS** logo; the **SOLIDWORKS Menus** will be displayed on the right of the logo. Note that the task pane is automatically closed when you start a new file and click in the drawing area. The initial screen display on starting a new part file of SOLIDWORKS using the **New** button in the **Menu Bar** is shown in Figure 2-4.

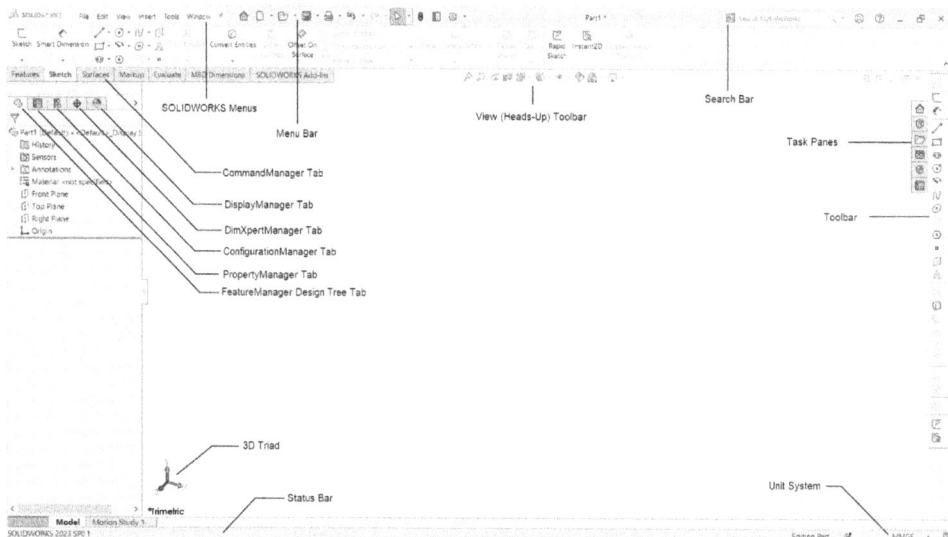

Figure 2-4 The components of a new part document

It is evident from the screen that SOLIDWORKS is a very user-friendly solid modeling software. Apart from the default CommandManager shown in Figure 2-4, you can also invoke other CommandManagers. To do so, right-click on the **CommandManager** tab; a shortcut menu will be displayed. Choose the required CommandManager from the shortcut menu; it will be added. Besides the existing CommandManager, you can also create a new CommandManager.

By default, **Flow Simulation CommandManager** is not displayed in SOLIDWORKS. To make it visible, choose the **SOLIDWORKS Flow Simulation** button from the **SOLIDWORKS Add-Ins CommandManager,** the **Flow Simulation CommandManager** will be displayed. Click on the **Flow Simulation CommandManager**, refer to Figure 2-5.

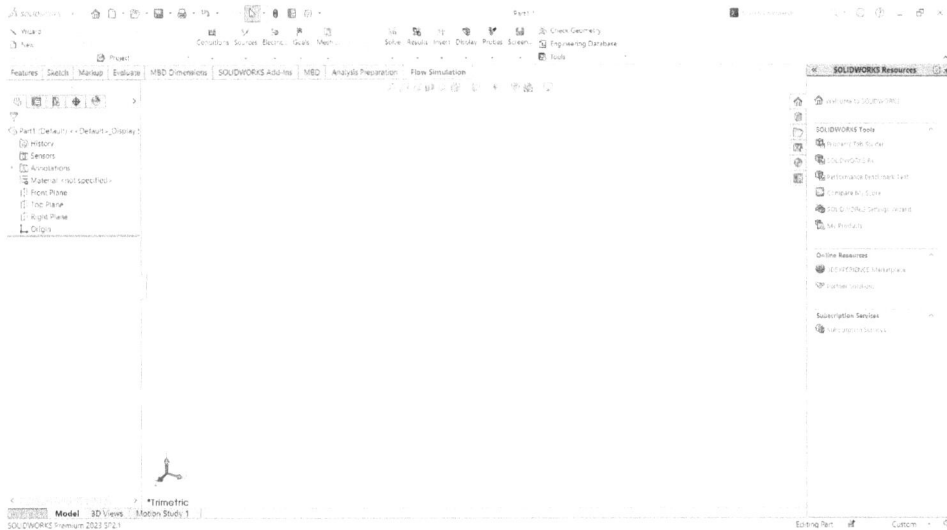

Figure 2-5 The Flow Simulation CommandManager

FLOW SIMULATION INTERFACE COMPONENTS

You need to create a project before understanding the interface components of the flow simulation. Next, you need to create a project by using the **Wizard** tool. To do so, choose the **Wizard** button; the **Wizard - Project Name** page is displayed, as shown in Figure 2-6.

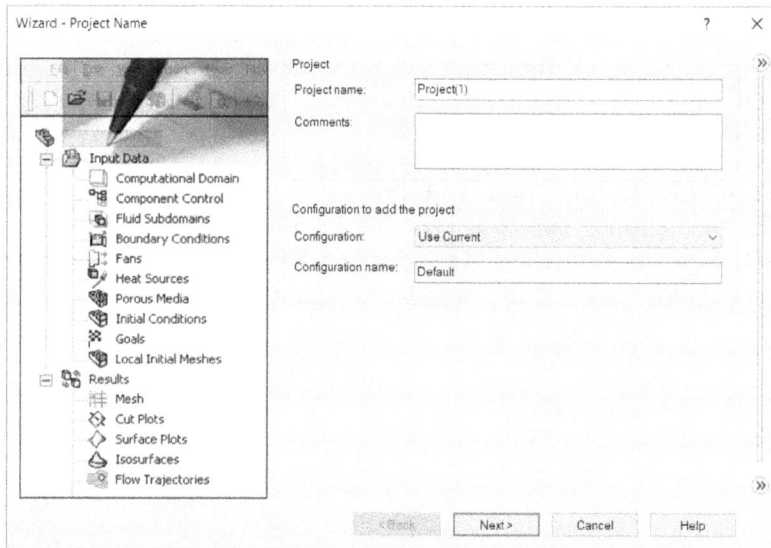

*Figure 2-6   The **Wizard - Project Name** page*

Enter the name **Flow around a sphere** in the **Project name** field. Choose the **Next** button; the **Wizard - Unit System** page is displayed, as shown in Figure 2-7. Select the **SI (m-kg-s)** system from the **Unit system** area if it is not selected by default.

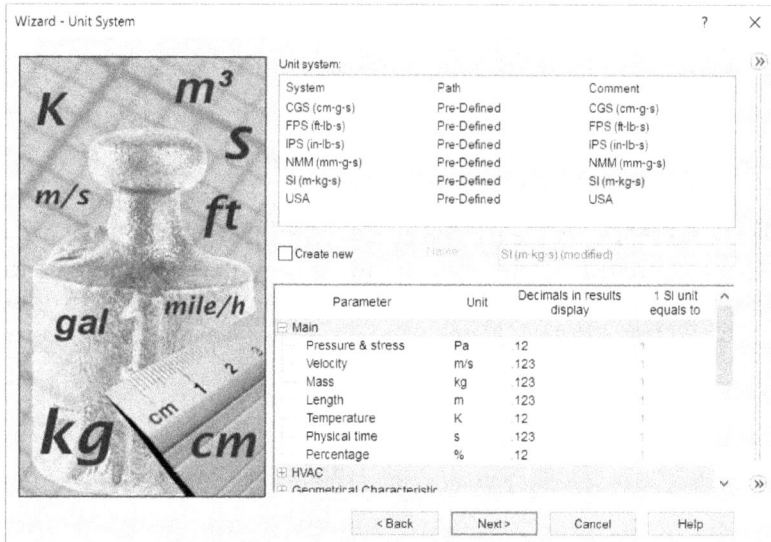

*Figure 2-7   The **Wizard - Unit System** page*

Choose the **Next** button; the **Wizard - Analysis Type** page is displayed, as shown in Figure 2-8.

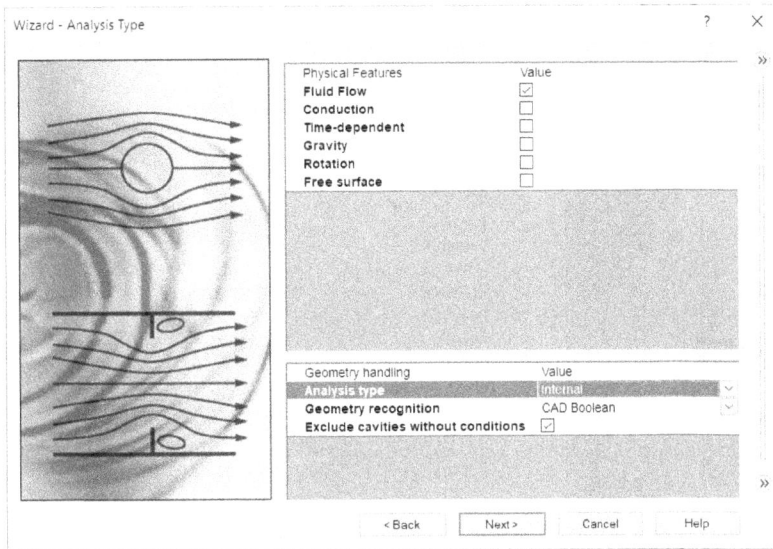

Figure 2-8 The **Wizard - Analysis Type** *page*

Select the **External** radio option under the **Analysis type** drop-down in the dialog box. Choose the **Next** button; the **Wizard - Default Fluid** page is displayed, as shown in Figure 2-9.

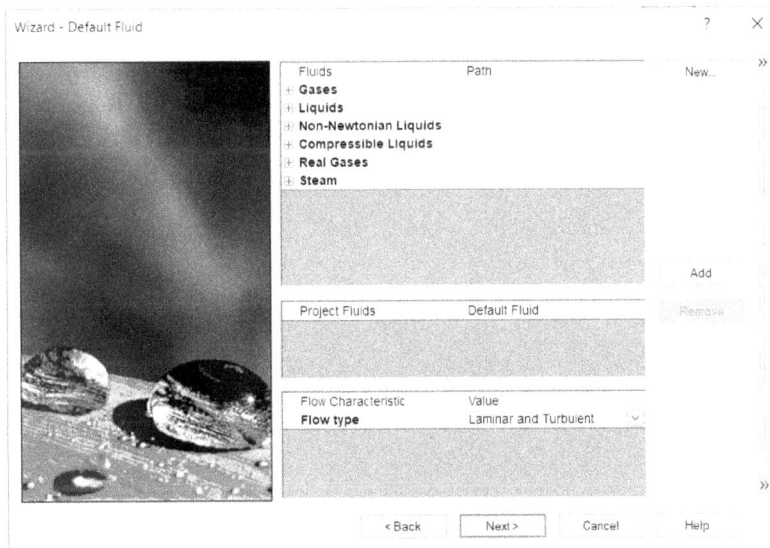

Figure 2-9 The **Wizard - Default Fluid** *page*

Click ⊞ sign on the left of the **Gases** option to display the list of fluids available under the **Gases category** in the **Fluids** column, as shown in Figure 2-10.

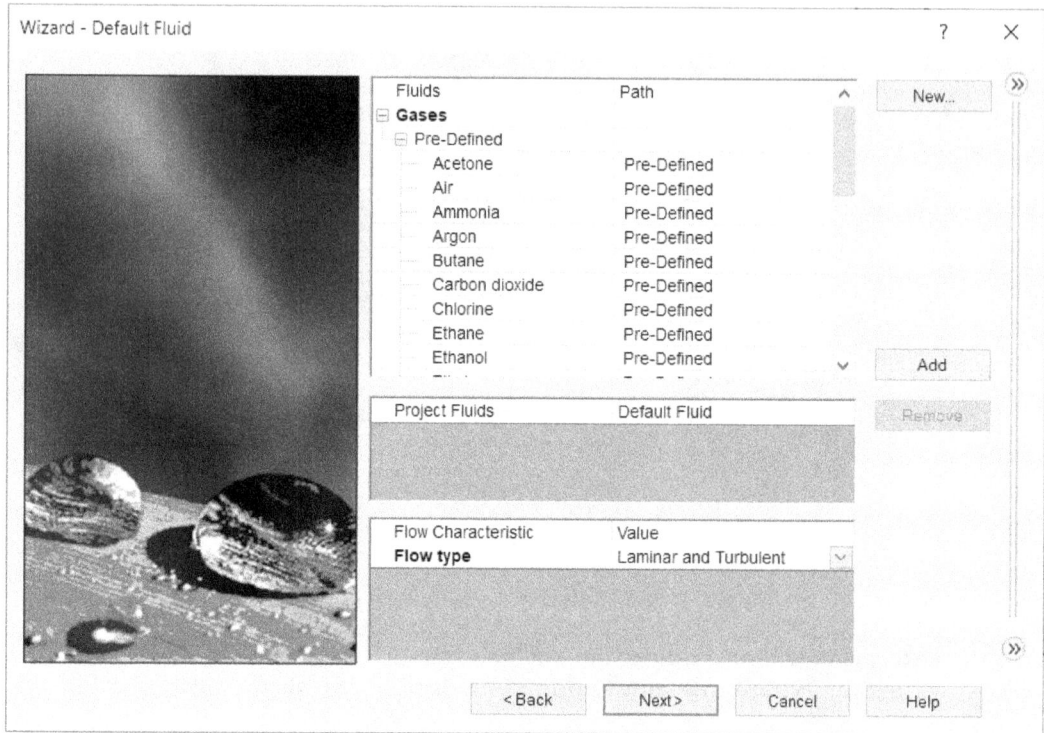

Figure 2-10 *The* ***Wizard - Default Fluid*** *page*

Select **Air** from the **Gases** column and choose the **Add** button; the **Air (Gases)** is added in the **Project Fluids** column. Choose the **Next** button; the **Wizard - Wall Conditions** page is displayed, as shown in Figure 2-11.

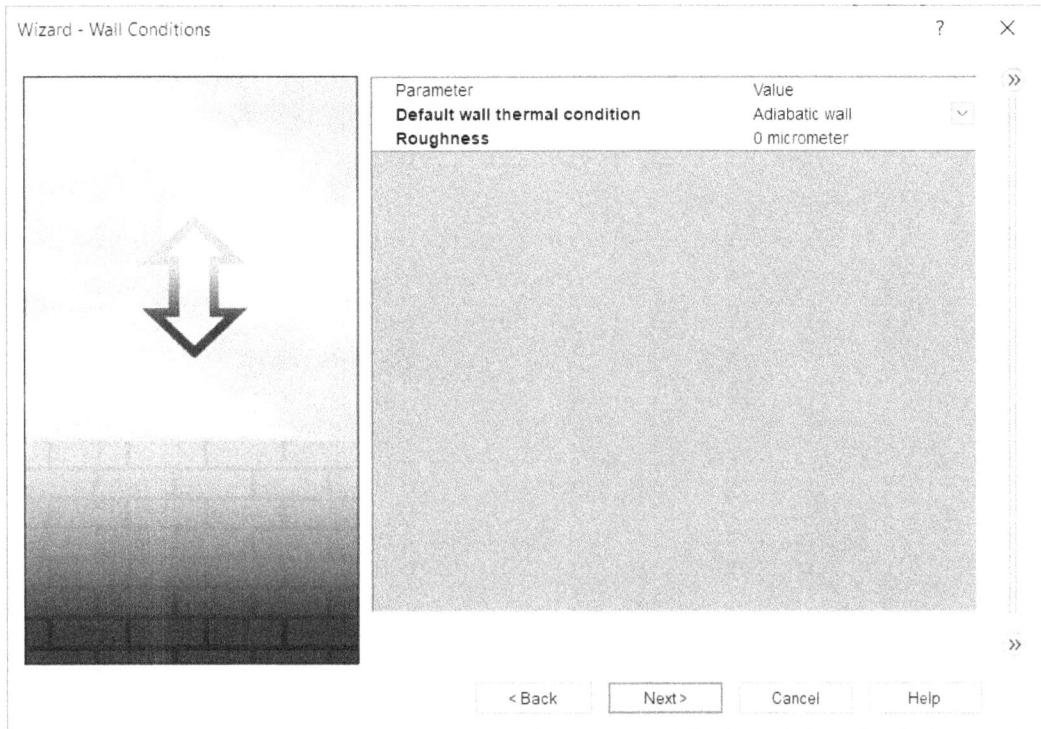

Figure 2-11 The *Wizard - Wall Conditions* page

Choose the **Next** button; the **Wizard - Initial Conditions** page is displayed, as shown in Figure 2-12.

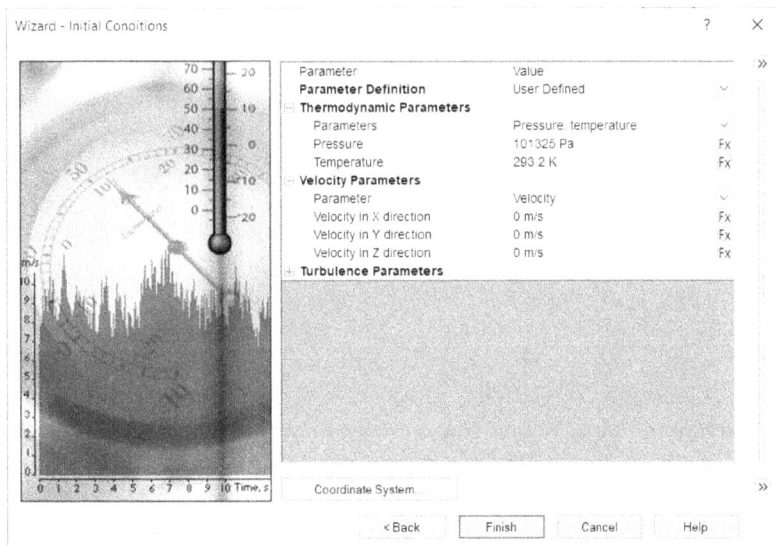

Figure 2-12 The *Wizard - Initial Conditions* page

Enter **0.006** in the **Velocity in X direction** edit box and choose the **Finish** button; a flow simulation project is created, as shown in Figure 2-13. A Flow Simulation project comprises all of a problem setups and outcomes. It includes both the physics and computational model setups, as well as any output data. It also contains simulation results such as a 3D velocity field or time-series pressure data. When you create a simple project, a new Flow Simulation Analysis tab opens on the right side of the FeatureManager design tree, PropertyManager, and ConfigurationManager.

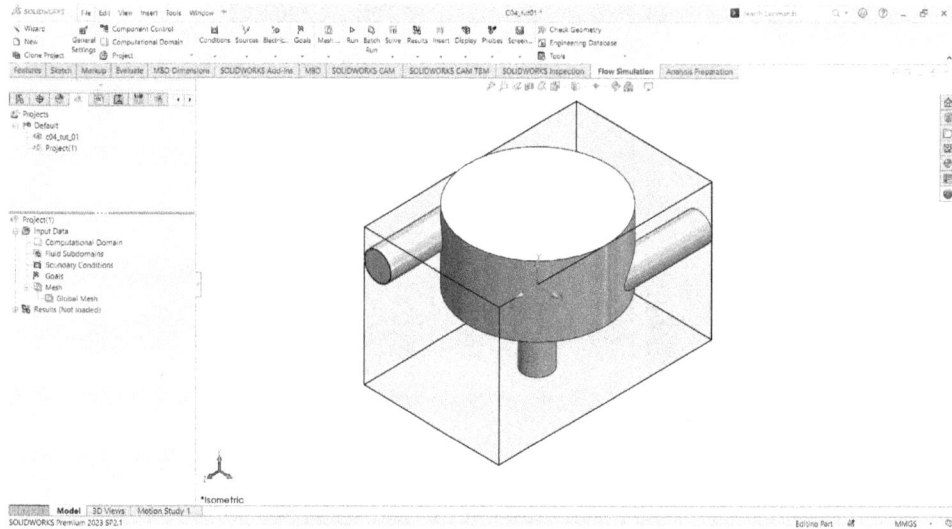

Figure 2-13 *A flow simulation project*

The flow simulation analysis tab has two panes. The upper pane of the Flow Simulation Analysis tab displays the Flow Simulation Projects Tree, which allows you to navigate from one project to the next and from one configuration to the next. In the lower pane of the tab, you can enter the remaining project data, such as boundary conditions, start circumstances, heat sources, solid materials, and goals.

FLOW SIMULATION MESHING IN SOLIDWORKS

SOLIDWORKS Flow Simulation uses a mesh of three-dimensional rectangular cells to describe fluid and solid volumes using the finite volume method. The mesh of cells that define the fluid and solid volumes is called a computational domain. You can view the finite volume mesh in the graphics area of SOLIDWORKS Flow Simulation. The computational domain is the region in which the mesh is generated.

Each cell in the computational domain has three sides and one or more walls. The top surface of each cell is assigned a zero-velocity condition. In addition to the boundaries of the computational domain, you can also create interfaces between two or more regions to simulate boundary conditions. It is essential to ensure that all walls and interfaces are consistent in either being two-dimensional or three-dimensional. In a given computational domain, all walls and interfaces must maintain uniform dimensionality.

In Flow Simulation, you can make use of two types of meshes, structured or unstructured mesh. Structured meshes are defined by three-dimensional rectangular cells, whereas unstructured meshes are defined by three-dimensional triangular cells. The advantage of structured meshes is that they are easier to handle and provide better stability and a more compact solution. On the other hand, unstructured meshes allow for a greater degree of freedom in modeling complex geometries such as curved surfaces or irregularly shaped volumes. Unstructured meshes can also be used for flow around solid objects with arbitrary shapes.

You can do the mesh settings globally and locally in the software. Select **Global Mesh** from the **Flow Simulation Design** tree and right-click on it; a shortcut menu is displayed. Select the **Edit Definition** option from the shortcut menu; the **Global Mesh Settings** PropertyManager is displayed, refer to Figure 2-14. You can select the **Automatic** or **Manual** button to specify the type of mesh from the **Type** rollout. By default, the **Automatic Mesh** button is selected in the **Type** rollout. In the **Settings** rollout, you can set the level of mesh by dragging the slider of level of initial mesh. When the number of level increases then more fine cells are produced but it will take greater CPU time and require more computer memory to solve the problem. You can enter the value in the **Minimum Gap Size** edit box to detect tiny geometries that Flow Simulation does not recognize automatically.

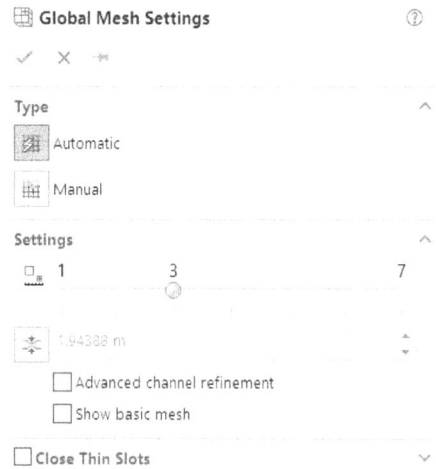

*Figure 2-14 The **Global Mesh Settings** PropertyManager*

When you select the **Manual** button from the **Type** rollout, then different rollouts with different options are displayed, refer to Figure 2-15. You can enter the value in the **Nx**, **Ny**, and **Nz** edit boxes to specify the numbers of basic mesh cells in the X, Y, and Z directions. Select the **Keep Aspect Ratio** check box from the **Basic Mesh** rollout to maintain the ratio between the numbers of basic mesh cells in each co-ordinate direction. Select the **Show** check box to see the generated basic mesh on the model. Choose the **OK** button after defining the settings in the dialog box.

Select the **Mesh** from the **Flow Simulation Design** tree and right-click on it; a shortcut menu is displayed. Select the **Insert Local Mesh** option from the shortcut menu; the **Local Mesh Settings** dialog box is displayed, refer to Figure 2-16. The settings in this dialog box enables you to provide an initial mesh in a specific region of the computational domain in order to successfully define model-specific geometry.

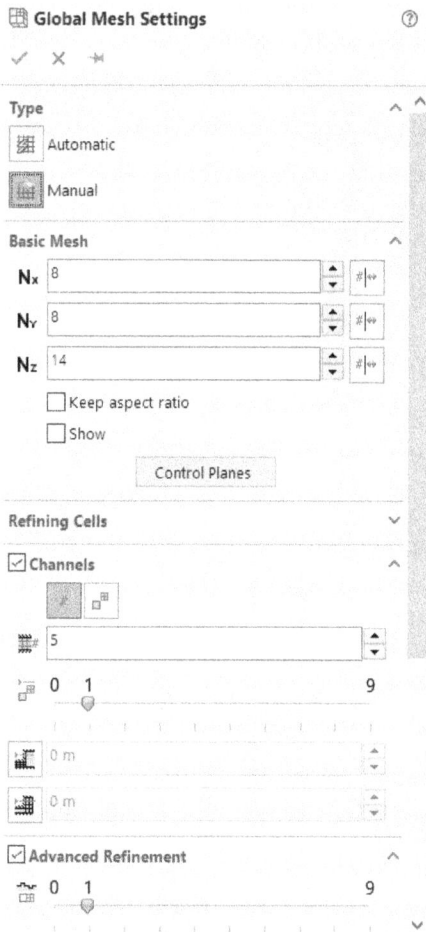

Figure 2-15 *The partial view of* **Global Mesh Settings PropertyManager** *with* **Manual** *button selected*

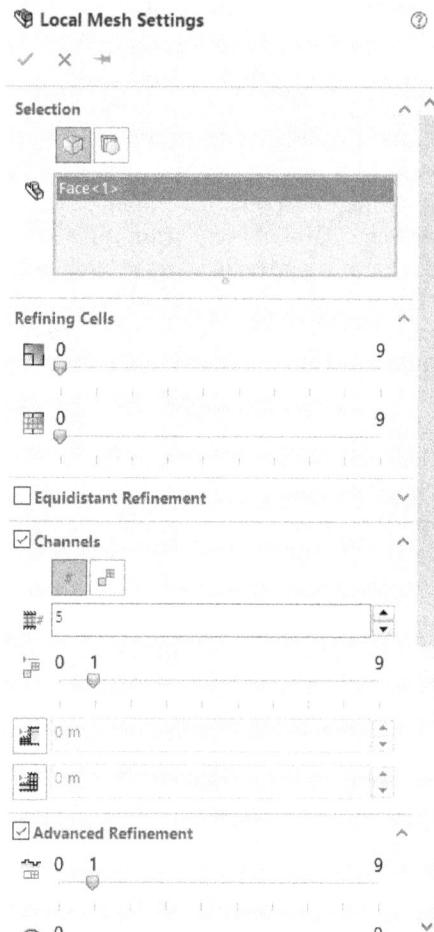

Figure 2-16 *The partial view of* **Local Mesh Settings PropertyManager**

CALCULATION CONTROL OPTIONS

You can set parameters that influence the Flow Simulation in the **Calculation Control Options** dialog box. Choose the **Calculation Control Options** tool from the **Solve** drop-down of the **Flow Simulation CommandManager**; the **Calculation Control Options** dialog box is displayed, refer to Figure 2-17. Following are the parameters which you can control by using the dialog box:

a)   Deciding whether or not to finish the calculation.

b)   In a time-dependent analysis, altering the time step.

c)   During the calculation, invoke the refining of the computational mesh.

d)   Adjusting the parameters of the radiation model on which the computation time and accuracy are based.

e)   Freezing values of all flow parameters.

f)   During the calculation, you can save the results.

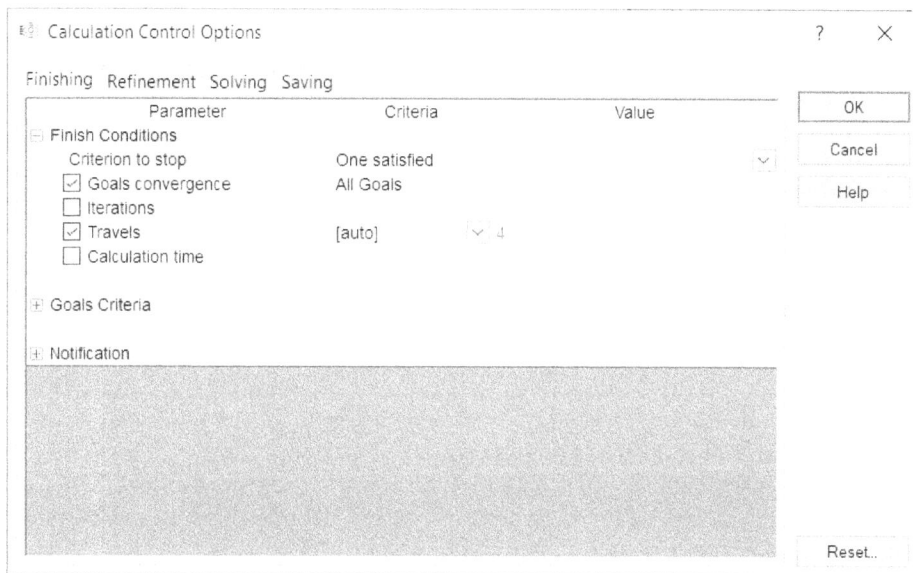

*Figure 2-17   The **Calculation Control Options** dialog box*

TOOLS

The Flow Simulation includes the following tools and they are discussed next:

a)   Create Lids

b)   Leak Tracking

c)   Calculator

d)   Copy to Project

e)   Options

These tools are discussed in detail next.

Create Lids

You can use this tool to create lids automatically. If you modify an opening, the tool will not automatically update it, so you will need to recreate it. To create the lids, choose the **Create Lids** tool from the **Tools** drop-down of the **Flow Simulation CommandManager**; the **Create Lids PropertyManager** is displayed, refer to Figure 2-18. Select the face from the model to close the opening; the dialog box gets modified, refer to Figure 2-19. Choose the **OK** button to close the dialog box; the lid will be added to the model.

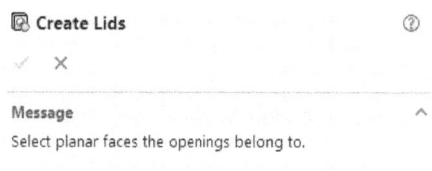

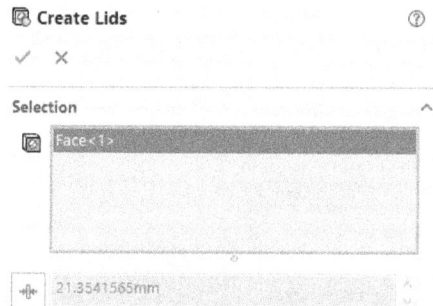

*Figure 2-18 The **Create Lids** PropertyManager*

*Figure 2-19 The modified **Create Lids** PropertyManager*

Leak Tracking

This tool is designed for identifying gaps and holes in your model. It is important to note that if you select Start and End faces that belong to unconnected components, the attachment path will not be established. Therefore, whenever possible, it is advisable to choose Start and End faces within the same component. To find the openings, choose the **Leak Tracking** tool from the **Tools** drop-down of the **Flow Simulation CommandManager**; the **Leak Tracking** tab is added at the lower left corner of the drawing area, refer to Figure 2-20. In the **Start Face** area, you need to select the inner or outer face of the component and in the **End Face** area, you need to select the outer or inner face of the component. Choose the **Find Connection** button to detect a gap or hole. The list of faces connected by the path can be found in the **Track Faces** list, refer to Figure 2-21.

*Figure 2-20 The **Leak Tracking** tab*

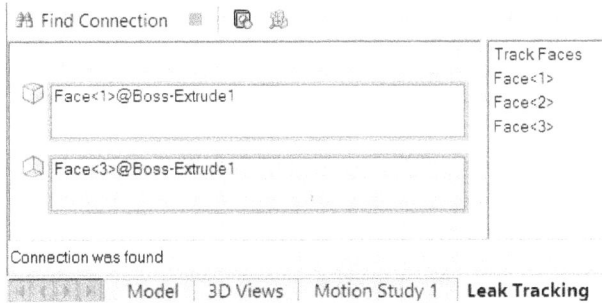

Figure 2-21 *The list of faces displayed in the* **Track Faces** *list*

Calculator

You can use the calculator to do a variety of manual calculations using engineering formulas. Choose the **calculator** tool from the **Tools** drop-down of the **Flow Simulation CommandManager**; the **Gasdynamic Calculator** window is displayed, refer to Figure 2-22. Choose the **New Formula** from the **Edit** menu; the **New Formula** dialog box is displayed. Select the required checkbox from the **Select the name of the new formula** area and choose the **OK** button. You will notice that the selected formula gets inserted into the **Gasdynamic Calculator** window. You can enter the values of the certain parameters in this window and then save the file. You can copy the data from the window to the Excel for further processing.

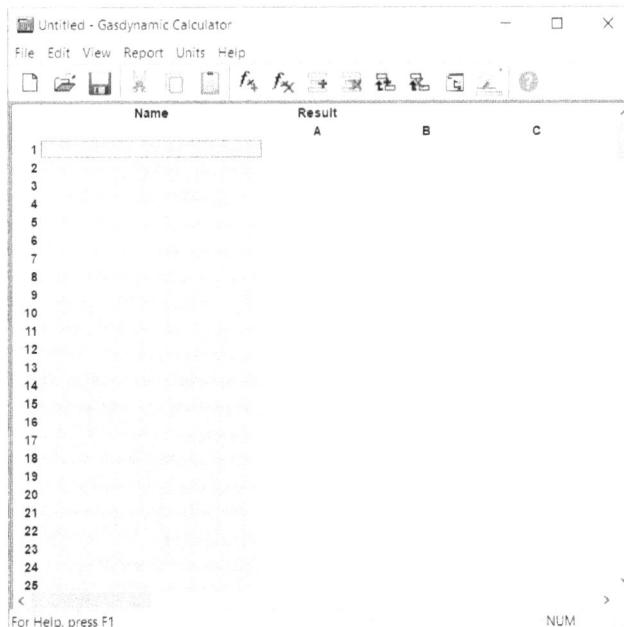

Figure 2-22 *The* **Gasdynamic Calculator** *dialog box*

Copy to Project

You can use the **Copy to Project** tool to copy the project's input data and results images to the current project. Choose the **Copy to Project** tool from the **Tools** drop-down of the **Flow Simulation CommandManager**; the **Copy to Project PropertyManager** is displayed, refer to Figure 2-23. Select features to copy from the drawing area; the selected features will be shown in the **Select features to Copy** area. Choose the **OK** button to close the dialog box.

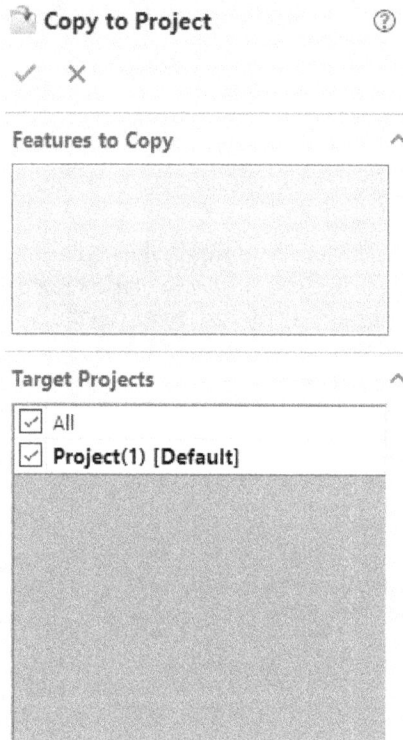

Figure 2-23 The *Copy to Project PropertyManager*

Options

You can use the **Options** tool to specify various settings for the Flow Simulation. Choose the **Options** tool from the **Tools** drop-down of the **Flow Simulation CommandManager**; the **Options** dialog box is displayed, refer to Figure 2-24. The parameters available in this dialog box are discussed next:

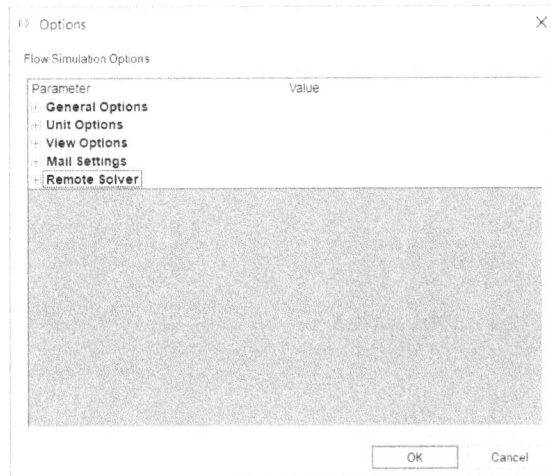

*Figure 2-24 The **Options** dialog box*

a) **General Options**: This option allows you to customize the various settings of the Flow Simulation.

b) **Unit Options**: This option allows you to change the default unit notation and define the default unit system.

c) **View Options**: This option allows you to select a variety of options related to the visualization feature.

d) **Mail Settings**: This option allows you to provide Flow Simulation with the information it needs to send e-mails.

e) **Remote Solver**: This option allows you to specify the data that the remote solver needs.

Self-Evaluation Test

Answer the following questions and then compare them to those given at the end of this chapter:

1. Which of the following button is available in the **Type** rollout of the **Global Mesh Settings PropertyManager**?

    (a) **Automatic**   (b) **Manual**
    (c) Both   (d) None of these

2. Which of the following PropertyManager is displayed when you choose the **Create Lids** tool from the **Flow Simulation CommandManager**?

    (a) **Automatic**   (b) **Manual**
    (c) Both   (d) None of these

3. You can do the mesh settings _____ and locally in the SOLIDWORKS software.

4. You can enter the value in the _____ edit box to detect tiny geometries that Flow Simulation does not recognize automatically.

Review Questions

Answer the following questions:

1. Which of the following button need to select from the **New SOLIDWORKS Document** dialog box to enter into the modeling environment?

   (a) **Assembly**   (b) **Part**
   (c) **Drawing**   (d) None of these

2. Which of the following radio button is available in the **Analysis** type area of **Wizard - Analysis Type** page?

   (a) **Internal**   (b) **External**
   (c) Both   (d) None of these

3. You can use the _____ to do a variety of manual calculations.

4. You can use the _____ tool to copy the project input data and results images to the current project.

Answers to Self-Evaluation Test
1. Both, 2. Both, 3. Globally, 4. Minimum Gap Size

Chapter 3

Creating and Preparing Model for Flow Simulation

Learning Objectives

After completing this chapter, you will be able to:

- *Create solid base extruded features*
- *Create thin base extruded features*
- *Create shell features*
- *Create sweep features*

BASIC TOOLS TO CREATE A MODEL

In SOLIDWORKS, there are various tools that can be used to create a model. Some of the tools which are needed to create a model are discussed in this chapter. For more tools in SOLIDWORKS, you can refer to SOLIDWORKS 2023 for Designers, 21st Edition authored by Prof. Sham Tickoo, Purdue University Northwest, USA. In this chapter, you will learn about the usage of the **Extrude** tool, which is used to extrude a sketch; the **Shell** tool, which is used to scoop out material from a model to make it hollow; and the **Sweep** tool, which is used to create a swept surface along a closed or open profile.

CREATING BASE FEATURES BY EXTRUDING SKETCHES

CommandManager:	Features > Extruded Boss/Base
SOLIDWORKS Menus:	Insert > Boss/Base > Extrude
Toolbar:	Features > Extruded Boss/Base > Extruded Boss/Base

The sketches that you have drawn can be converted into base features by extruding the sketch using the **Extruded Boss/Base** tool from the **Features CommandManager**. After drawing the sketch, choose the **Features** tab from the **CommandManager** to display the **Features CommandManager**. Next, choose the **Extruded Boss/Base** tool from the **Features CommandManager**; the sketching environment will be closed and the part modeling environment will be invoked. Also, the preview of the feature that is created using the default options will be displayed in the trimetric view. The trimetric view gives a better display of the solid feature.

On the basis of the options and the sketch selected for extruding, the resulting feature can be a solid feature or a thin feature. If the sketch is closed, it can be converted into a solid feature or a thin feature. However, if the sketch is open, it will be converted into a thin feature. The process of creating a solid or thin feature is discussed next.

Creating Solid Extruded Features

After you have completed drawing a closed sketch, dimension it to convert it into a fully defined sketch. Next, choose the **Features** tab from the **CommandManager**; the **Features CommandManager** will be displayed. Choose the **Extruded Boss/Base** tool; the **Boss-Extrude PropertyManager** will be displayed, refer to Figure 3-1. Also, you will notice that the view is automatically changed to the trimetric view.

You will also notice that the preview of the base feature is displayed in temporary graphics. Additionally, an arrow will appear in front of the sketch. Note that if the sketch consists of some closed loops inside the outer loop, they will automatically be subtracted from the outer loop while extruding, as shown in Figure 3-2.

The options in the **Boss-Extrude PropertyManager** are discussed next.

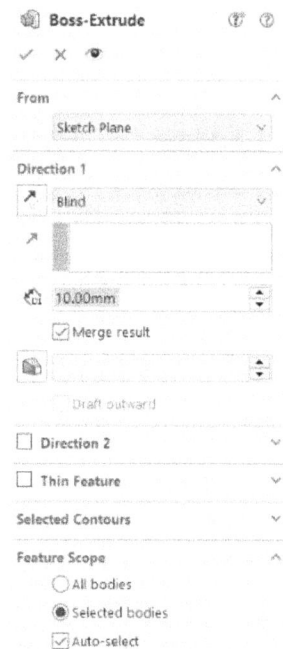

Figure 3-1 The Boss-Extrude PropertyManager

Direction 1

The **Direction 1** rollout is used to specify the end condition for extruding the sketch in one direction from the sketch plane. The options in the **Direction 1** rollout list are discussed next.

End Condition

The **End Condition** drop-down list provides options to define the termination of the extruded feature. Note that when you create the first feature, some of the options in this drop-down list will not be used. Also, some additional options will be available later in this drop-down list. Some of the options that will be used to define the termination of the base feature are discussed next.

Blind

The **Blind** option is selected by default and is used to define the termination of the extruded base feature by specifying the depth of extrusion. The depth of the extrusion is specified in the **Depth** spinner. This spinner will be displayed in the **Direction 1** rollout on selecting the **Blind** option. To reverse the extrusion direction, choose the **Reverse Direction** button provided on the left of this drop-down list. Figure 3-2 shows the preview of the feature being created by extruding the sketch using this option.

You can also extrude a sketch to a blind depth by dragging the feature dynamically using the mouse. To do so, move the cursor to the arrow displayed in the preview; the move cursor will be displayed and the color of the arrow will also change. Left-click once on the arrow; a scale will be displayed, as shown in Figure 3-3. Now, move the cursor to specify the depth of extrusion; the value of the depth of extrusion will change dynamically on this scale as you move the cursor. Left-click again to specify the termination of the extruded feature; the select cursor will be replaced by the mouse cursor. Right-click and choose **OK** to complete the feature creation or choose the **OK** button from the **Boss-Extrude PropertyManager**.

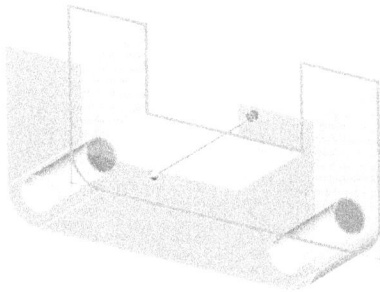

Figure 3-2 *Preview of the feature being extruded using the **Blind** option*

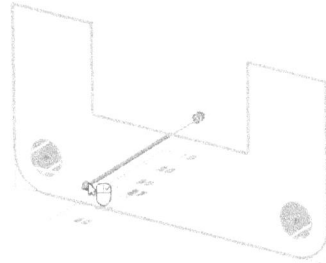

Figure 3-3 *Preview of the feature being extruded by dragging the arrow dynamically*

Mid Plane

The **Mid Plane** option is used to create the base feature by extruding the sketch equally in both the directions of the plane on which the sketch is drawn. For example, if the total depth of the extruded feature is 30 mm, it will be extruded 15 mm toward the front of the sketching plane and 15 mm toward the back. The depth of the feature can be defined in the **Depth** spinner that is displayed below the **Direction of Extrusion** selection box. Figure 3-4 shows preview of the feature being created by extruding the sketch using the **Mid Plane** option.

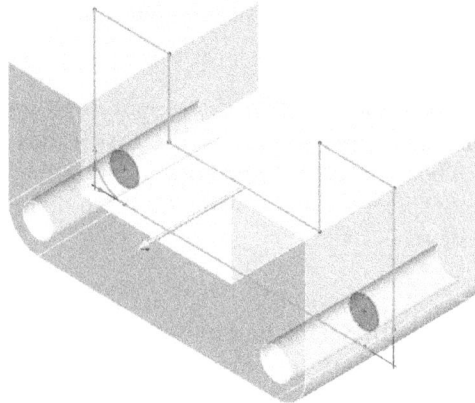

Figure 3-4 *Preview of the feature being extruded using the **Mid Plane** option*

Draft On/Off

The Draft On/Off button is used to specify a draft angle while extruding a sketch to taper the resulting feature. To add a draft angle to the feature, choose this button; the Draft Angle spinner and the Draft outward check box will be available. You can enter the draft angle for the feature in the Draft Angle spinner. By default, the feature will be tapered inward, as shown in Figure 3-5.

If you want to taper the feature outward, select the Draft outward check box that is displayed below the Draft Angle spinner. The feature created with the outward draft is shown in Figure 3-6.

Note
The Direction of Extrusion area will be discussed in the later chapters.

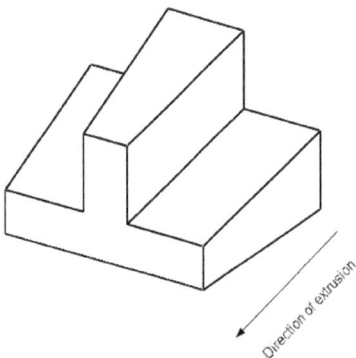

Figure 3-5 *Feature created with inward draft*

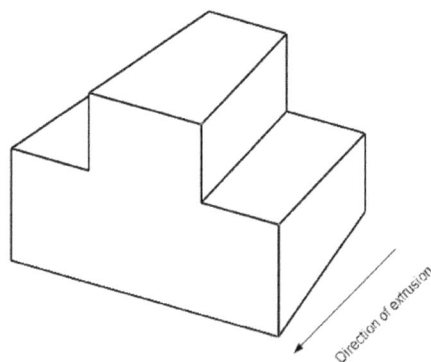

Figure 3-6 *Feature created with outward draft*

Direction 2

The **Direction 2** rollout is used to extrude a sketch with different values in the second direction of the sketching plane. The **Direction 2** rollout will be activated only when you select the **Direction 2** check box.

The options in this rollout are similar to those in the **Direction 1** rollout. Note that unlike the **Mid Plane** termination option, the depth of extrusion and other parameters in both directions can be different. For example, you can extrude the sketch to a blind depth of 20 mm and an inward draft of 15-degree in front of the sketching plane, and to a blind depth of 15 mm and an outward draft of 0-degree behind the sketching plane, as shown in Figure 3-7.

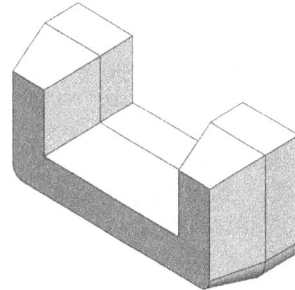

Figure 3-7 Feature created in two

After setting the values for both directions, choose the **OK** button or choose the **OK** icon from the confirmation corner; the feature will be created with defined values.

directions with different values

Creating Thin Extruded Features

The thin extruded features can be created using a closed or an open sketch. If the sketch is closed, the thickness will be specified inside or outside the sketch to create a cavity inside the feature, as shown in Figure 3-8. To convert a closed sketch into a thin feature, select the check box in the **Thin Feature** rollout title bar; the rollout will expand, as shown in Figure 3-9.

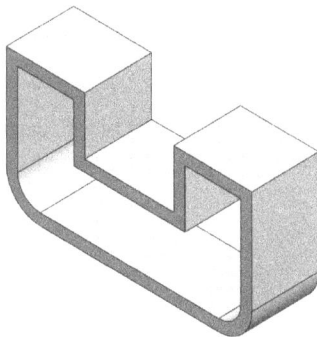

Figure 3-8 Thin extruded feature created using a closed loop

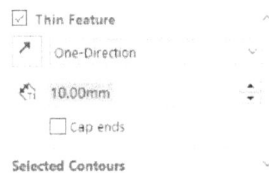

Figure 3-9 Feature created in two directions with different values

The options in the **Thin Feature** rollout of the **Boss-Extrude PropertyManager** are discussed next.

Type

The options provided in the **Type** drop-down list are used to select the method of defining the thickness of the thin feature. These options are discussed next.

One-Direction

The **One-Direction** option is used to add thickness on one side of the sketch. The amount of thickness to be applied can be specified in the **Thickness** spinner provided below the **Type** drop-down list. For the closed sketches, the direction can be inside or outside the sketch. Similarly, for open sketches, the direction can be below or above the sketch. You can reverse the direction of thickness using the **Reverse Direction** button available on the left of the **Type** drop-down list. This button will be available only when you select the **One-Direction** option from this drop-down list.

Mid-Plane

The **Mid-Plane** option is used to add thickness equally on both sides of the sketch. The value of the thickness of the thin feature can be specified in the **Thickness** spinner provided below this drop-down list.

Two-Direction

The **Two-Direction** option is used to create a thin feature by adding different thicknesses on both sides of a sketch. The thickness values in direction 1 and direction 2 can be specified in the **Direction 1 Thickness** spinner and the **Direction 2 Thickness** spinner, respectively. These spinners will automatically be displayed below the **Type** drop-down list when you select the **Two-Direction** option from this drop-down list.

Cap ends

The **Cap ends** check box will be displayed only when you select a closed sketch for converting into a thin feature. This check box is selected to cap the two open faces of the thin extruded feature. Both the open faces will be capped with a face having specified thickness. When you select this check box, the **Cap Thickness** spinner will be displayed below this check box. The thickness of the end caps can be specified by using this spinner.

If the sketch to be extruded is open, as shown in Figure 3-10, the **Thin Feature** rollout will be invoked automatically on invoking the **Boss-Extrude PropertyManager**. The resulting feature is shown in Figure 3-11.

Figure 3-10 *Open sketch to be converted into a thin feature*

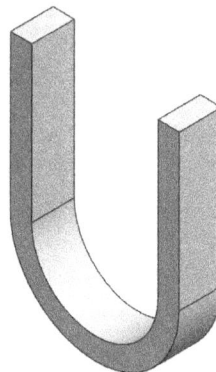

Figure 3-11 *Resulting thin feature*

Auto-fillet corners

The **Auto-fillet corners** check box will be displayed only when you select an open sketch to convert it into a thin feature. If you select this check box, all sharp vertices in the sketch will automatically be filleted during conversion into a thin feature. As a result, the thin feature will have filleted edges. The radius of the fillet can be specified in the **Fillet Radius** spinner, which will be displayed below the **Auto-fillet corners** check box.

Figure 3-12 shows the thin feature created by extruding an open sketch in both directions. Note that a draft angle is applied to the feature while extruding in the front direction and the **Auto-fillet corners** check box is selected while creating this thin feature.

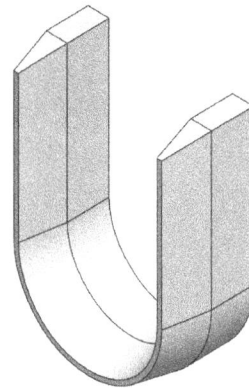

Figure 3-12 Thin feature created in both directions

Creating Shell Features

CommandManager: Features > Shell
SOLIDWORKS menus: Insert > Features > Shell
Toolbar: Features > Shell

Shelling is a process in which the material is scooped out from a model. The resulting model will be a hollow model with walls of a specified thickness and a cavity inside. The selected face or the faces of the model are also removed in this operation. If you do not select a face to be removed, a closed hollow model will be created. You can also specify multiple thicknesses to the walls.

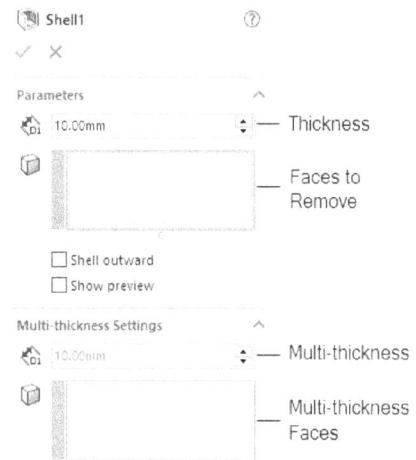

To create a shell feature, invoke the **Shell** tool from the **Features CommandManager**; the **Shell PropertyManager** will be displayed, as shown in Figure 3-13. Also, you will be prompted to select the faces to be removed. Select the face or the faces of the model that you want to remove. The selected faces will be highlighted in blue and their names will be displayed in the **Faces to Remove** selection box. Set

Figure 3-13 *The Shell PropertyManager*

the value of the wall thickness in the **Thickness** spinner and choose the **OK** button from the **Shell PropertyManager**. Figure 3-14 shows the face selected to remove and Figure 3-15 shows the resultant shell feature.

If none of the faces are selected to be removed, the resulting model will be hollowed from inside with no face removed. Figure 3-16 shows a model in the **Hidden Lines Visible** mode with a shell feature in which no face is selected to be removed.

Figure 3-14 *Face to be removed*

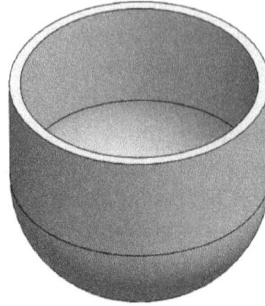

Figure 3-15 *Resultant shell feature*

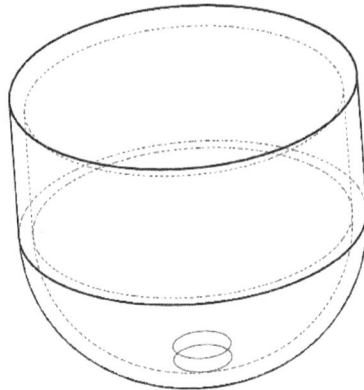

Figure 3-16 *Shell feature with no face selected to be removed*

Based on the geometric conditions, the quantity of the material to be removed from the shell feature will be decided automatically. Figure 3-17 shows the shell feature whose wall thickness is small for uniform shelling of the entire model. Figure 3-18 shows the shell feature whose wall thickness is large. As a result, it cannot accommodate uniform shelling of the entire model. Therefore, the material will not be removed from the area where the material removal is not possible.

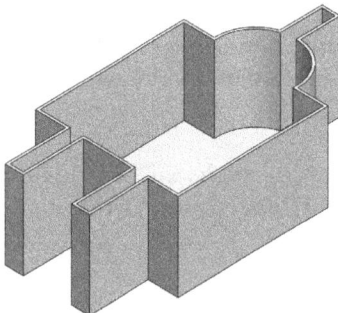

Figure 3-17 *Shell feature with small shell thickness*

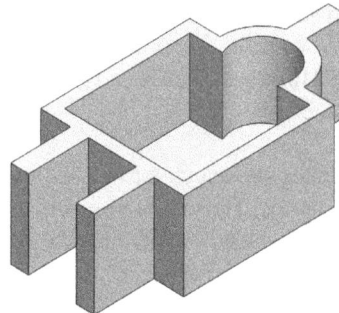

Figure 3-18 *Shell feature with larger shell thickness*

The **Shell outward** check box is selected to create the shell feature on the outer side of the model. You can also display the preview of the shell feature by selecting the **Show preview** check box from the **Parameters** rollout of the **Shell PropertyManager**.

Creating Multi-thickness Shell

The **Shell** tool can be used to shell the model by applying different thickness values to the selected faces. To do so, invoke the **Shell PropertyManager**, select the faces to be removed and then specify the thickness in the **Thickness** spinner of the **Parameters** rollout. Click once in the **Multi-thickness Faces** selection box to activate the selection mode. Select the faces for which you want to specify different thicknesses. Set the thickness value using the **Multi-thickness(es)** spinner and choose the **OK** button. Note that you can specify different thickness for each face. Figure 3-19 shows the faces selected to create a multi-thickness shell and Figure 3-20 shows the resulting shell feature.

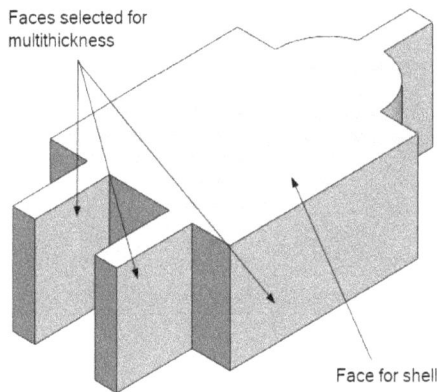

Faces selected for
multithickness

Face for shell

Figure 3-19 *Faces selected to create the multi-thickness shell feature*

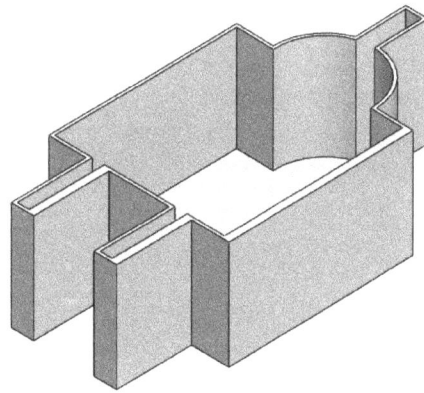

Figure 3-20 *Shell feature with larger shell thickness*

Error Diagnostics

While creating the shell feature, specify all parameters in the **Shell PropertyManager** and choose the **OK** button. If the creation of shell feature fails because of geometric inconsistency, the **Error Diagnostics** rollout will be displayed in the **Shell PropertyManager**, as shown in Figure 3-21. Also, the **Rebuild Errors** dialog box will be displayed informing you about the possible errors that can lead to the failure of feature creation. You will learn more about the **Rebuild Errors** dialog box in the later chapters. In the **Error Diagnostics** rollout, you can figure out the possible reasons behind the failure of the shell feature creation.

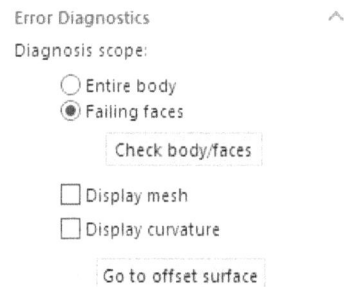

Error Diagnostics

Diagnosis scope:

○ Entire body
◉ Failing faces

Check body/faces

☐ Display mesh
☐ Display curvature

Go to offset surface

Figure 3-21 *The **Error Diagnostics** rollout*

The radio buttons in the **Diagnosis scope** area are used to specify whether the diagnosis has to be done on the entire body or only on the faces that have failed while being shelled. The **Check body/faces** button is used to run the diagnostic tool. On choosing this button, the areas of the model that can lead to failure of feature creation will be highlighted using callouts. The

Display mesh check box is used to display the surface curvature mesh. The **Display curvature** check box is selected to display the surface curvature. The **Go to offset surface** button is used to open **Offset Surface PropertyManager** where you can offset faces with smaller gaps which sometimes lead to errors.

Creating Sweep Features

CommandManager:	Features > Swept Boss/Base
SOLIDWORKS menus:	Insert > Boss/Base > Sweep
Toolbar:	Features > Extruded Boss/Base > Swept Boss/Base

One of the most important advanced modeling tools is the **Swept Boss/Base** tool. This tool is used to extrude a closed profile along an open or a closed path. Therefore, you need a profile and a path to create a sweep feature. A profile is a section for the sweep feature and a path is the course taken by the profile while creating the sweep feature. The profile has to be a sketch, but the path can be a sketch, curve, or an edge. You will learn more about the procedure to create the curves later in this chapter. Figure 3-22 shows a profile and a path for creating a sweep feature.

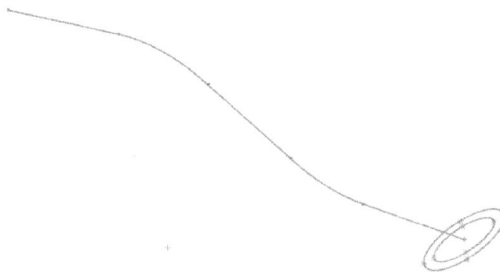

Figure 3-22 Profile and path to create a sweep feature

To create a sweep feature, choose the **Swept Boss/Base** button from the **Features** CommandManager; the **Sweep PropertyManager** will be displayed, as shown in Figure 3-23. You can create swept feature using sketch based profile or circular profile. By default, the **Sketch Profile** radio button is selected. As a result, you are prompted to select a sweep profile. Select the sketch drawn as the profile from the drawing area; the sketch will be highlighted and the **Profile** callout will be displayed. Also, you will be prompted to select a path for the sweep feature. Select the sketch or an edge to be used as the path; selection will be highlighted in magenta and the **Path** callout will be displayed. Also, the preview of the sweep feature will be displayed in the drawing area. Choose the **OK** button from the **Sweep PropertyManager** to end the feature creation. Figure 3-24 shows the resulting sweep feature.

To create a sweep feature using a circular profile along a sketch line, edge, or curve directly on a model, you need to choose

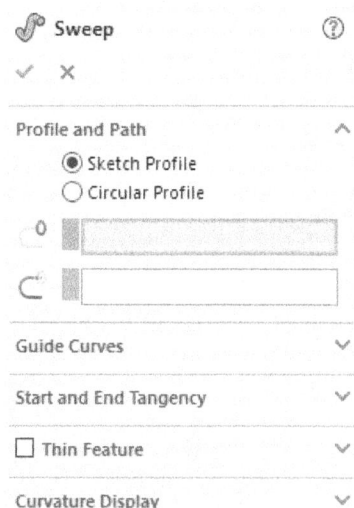

Figure 3-23 The Sweep PropertyManager

the **Circular Profile** button from the **Profile and Path** rollout in the **Sweep PropertyManager**. Next, from the drawing area, select the sketch or edge as a path along which the circular profile is to be created from the drawing area and enter the diameter value in the **Diameter** spinner.

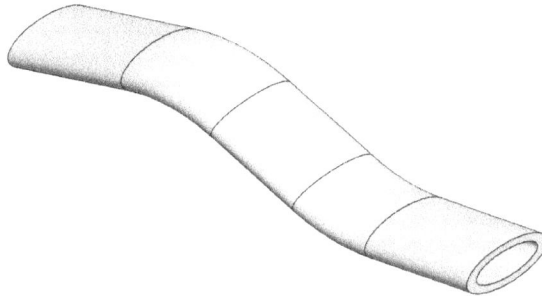

Figure 3-24 Sweep feature

TUTORIALS

Tutorial 1

In this Tutorial, you will create a model and make it ready for flow simulation. The model and its drawing are shown in Figure 3-25 and 3-26. The shell thickness of the model is 10 mm.

(Expected time: 30 min)

The following steps are required to complete this tutorial:

a. Start SOLIDWORKS and then start a new part document.
b. Create the base feature.
c. Create Inlet 1 in the base feature.
d. Create Inlet 2 in the base feature.
e. Create Outlet in the base feature.
f. Shell the feature for fluid flow.
g. Create Lid 1, Lid 2, and Lid 3.

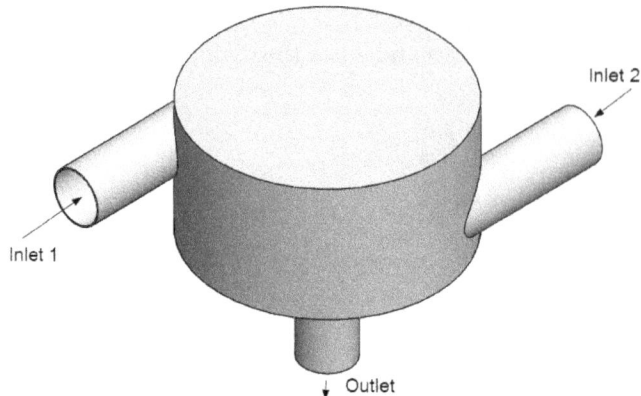

Figure 3-25 *Solid model for Tutorial 1*

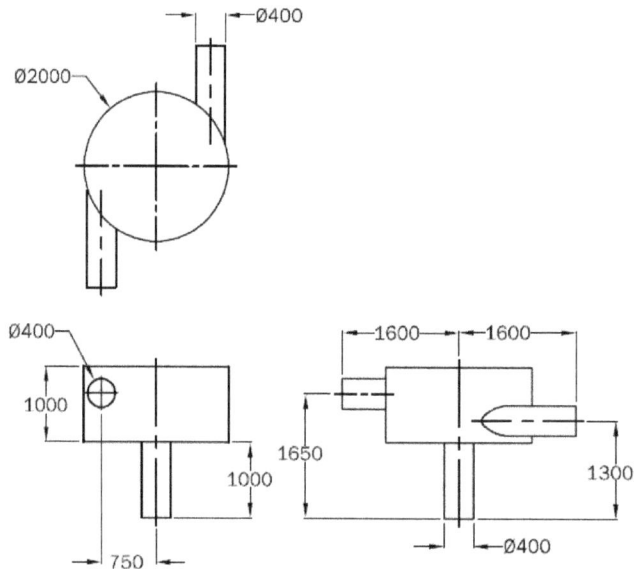

Figure 3-26 *Sketch for Tutorial 1*

Starting SOLIDWORKS and then a New SOLIDWORKS Document

1. Start SOLIDWORKS by double-clicking on the shortcut icon of SOLIDWORKS 2023 on the desktop of your computer; the SOLIDWORKS 2023 window along with the **Welcome - SOLIDWORKS** dialog box is displayed.

2. Choose the **Part** button available in the **New** area of the **Home** tab in the **Welcome - SOLIDWORKS** dialog box. A new **SOLIDWORKS** part document starts. Next, you need to invoke the sketching environment.

3. Choose the **Sketch** tab from the **CommandManager** if not chosen by default. Next, choose the **Sketch** button from the **Sketch CommandManager**; the **Edit Sketch PropertyManager** is invoked and you are prompted to select the plane to create the sketch.

4. Select the **Top Plane** from the drawing area; the sketching environment is invoked and the plane gets oriented normal to the view. You will notice that red colored arrows are displayed at the center of the screen indicating that you are in the sketching environment. Also, the confirmation corner with the Exit Sketch and Cancel options is displayed on the upper right corner in the graphics area. The screen display in the sketching environment of SOLIDWORKS 2023 is shown in Figure 3-27.

Figure 3-27 *Screen display in the sketching environment*

Drawing the Sketch

Next, you need to draw the sketch using the sketch tools. As it is evident from Figure 3-26, the sketch needs to be drawn using the **Circle** tool.

1. Press and hold the right-mouse button and drag the cursor to the right; the mouse gesture is displayed with sketching tools. Move the cursor over the **Circle** tool; the **Circle** tool is invoked and the **Circle PropertyManager** is displayed.

2. Make sure that the **Circle** button is chosen in the **Circle Type** rollout in the **Circle PropertyManager**. Move the circle cursor to the origin and click when an orange circle is displayed to specify the center point of the circle.

3. Move the cursor horizontally toward the right and draw a circle close to **2000** mm diameter and exit the **Circle** tool.

Next, add dimension to the sketch.

4. Choose the **Smart Dimension** tool and select the circle; the dimension is attached to the cursor. Place the dimension; the **Modify** dialog box is displayed with the default value. Enter **2000** in the **Modify** dialog box and press **Enter**, refer to Figure 3-28.

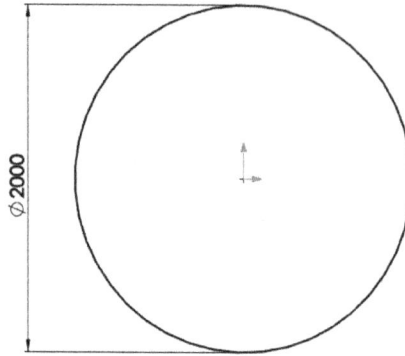

Figure 3-28 *Sketch after applying dimension*

Extruding the Sketch

Next, you need to invoke the **Extruded Boss/Base** tool and extrude the sketch using the parameters given in the tutorial description.

1. Choose the **Features** tab from the CommandManager to display the **Features CommandManager**. Then, choose the **Extruded Boss/Base** button; the sketch is automatically oriented to the trimetric view and the **Boss-Extrude PropertyManager** is displayed, as shown in Figure 3-29.

    As you are converting the closed sketch into a feature, only the **Direction 1** rollout is displayed in the **Boss-Extrude PropertyManager**. Also, a preview of the feature is displayed in the temporary shaded graphics with the default values.

2. Set **1000** in the **Depth** spinner as the depth in the **Direction 1** rollout and then choose the **OK** button; the extrude feature is created, as shown in Figure 3-30.

Boss-Extrude ⓘ

✓ ✕ ●

From ⌃

Sketch Plane ⌄

Direction 1 ⌃

↗ Blind ⌄

↗ [   ]

⚅ D1 10.00mm ⬍

⬛ [   ] ⬍

Draft outward

☐ Direction 2 ⌄

☐ Thin Feature ⌄

Selected Contours ⌄

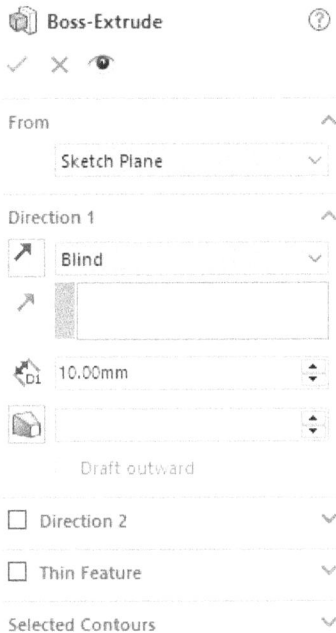

Figure 3-29 The **Boss Extrude PropertyManager**   *Figure 3-30* Sketch for Tutorial 1

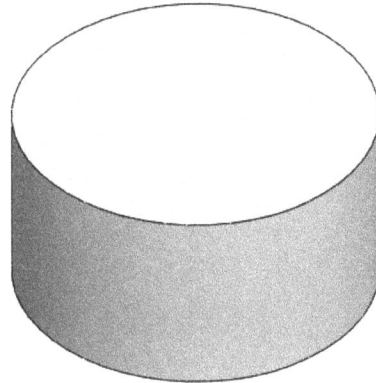

Creating the Inlet Pipe 1

Now, you need to create a cylindrical feature to get the required shape of the base feature. The sketch for this feature is created using a reference plane that is offset from Front plane.

1. Choose the **Plane** tool from the **Reference Geometry** flyout in the **Features CommandManager** to display the **Plane PropertyManager**.

2. Select the Front plane as the first reference and choose the **Offset distance** button from the **First Reference** rollout if not selected by default; the **Offset distance** spinner, and the **Flip offset** check box, are displayed in the **Plane PropertyManager**.

3. Set the value in the **Offset distance** spinner to **1600** and choose the **OK** button from the **Plane PropertyManager**; the required plane is created, as shown in Figure 3-31.

4. Select the reference plane which you just created if not already selected, and invoke the sketching environment. Set the current view normal to the eye view.

5. Draw the sketch of the circle and apply required relations to the sketch, as shown in Figure 3-32.

6. Change the current view to the isometric view and invoke the **Extruded Boss/Base** tool. You will observe in the preview that the direction of the feature creation is opposite to the required direction. Therefore, you need to change the direction of the feature creation.

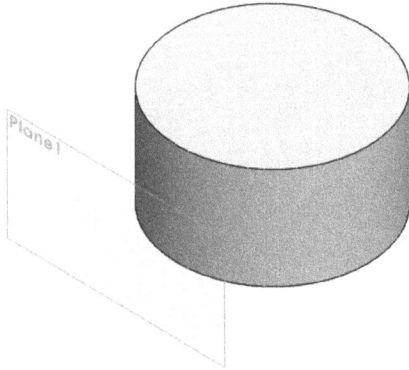

Figure 3-31 *The plane created*

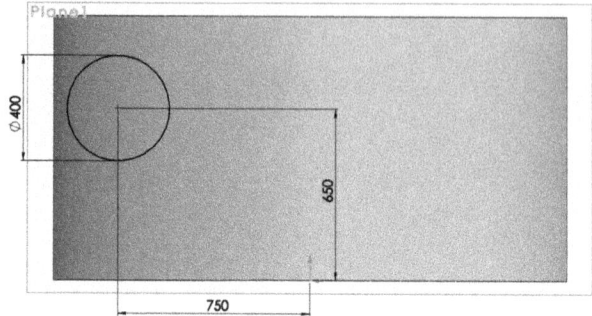

Figure 3-32 *Sketch created on plane*

7. Choose the **Reverse Direction** button on the left of the **End Condition** drop-down list to reverse the direction of feature creation; preview of the feature changes dynamically.

8. Right-click in the drawing area and choose the **Up To Surface** option from the shortcut menu; you are prompted to select a face or a surface to specify the first direction. Also, the **Face/Plane** selection box is displayed below the **Direction of Extrusion** selection box in the **Direction 1** rollout.

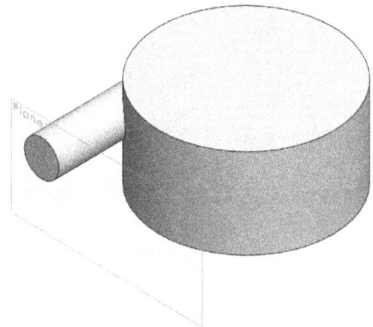

9. Select the curved surface of the model using the left mouse button. You will observe in the preview that the feature is extruded up to the selected surface, as shown in Figure 3-33.

Figure 3-33 *Inlet Pipe 1 created*

10. Choose the **OK** button from the **Boss-Extrude PropertyManager**.

    For better display turn off the plane display.

Creating the Inlet Pipe 2

Now, you need to create a cylindrical feature to get required shape of the base feature. The sketch for this feature is created using a reference plane that is offset from Front plane.

1. Choose the **Plane** tool from the **Reference Geometry** flyout in the **Features CommandManager** to display the **Plane PropertyManager**.

2. Select the Front plane as the first reference and choose the **Offset distance** button from the **First Reference** rollout if not selected by default; the **Offset distance** spinner and the **Flip offset** check box are displayed in the **Plane PropertyManager**.

3. Set the value in the **Offset distance** spinner to **1600** and choose the **OK** button from the **Plane PropertyManager**; the required plane is created. Select the **Flip offset** check box to reverse the direction of plane, as shown in Figure 3-34.

4. Select the reference plane which you just created if not already selected, and invoke the sketching environment. Set the current view normal to the eye view.

5. Draw the sketch of the circle and apply required relations and dimensions to the sketch, as shown in Figure 3-35.

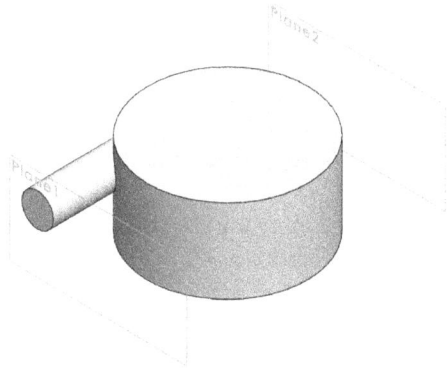

Figure 3-34 *The plane created*

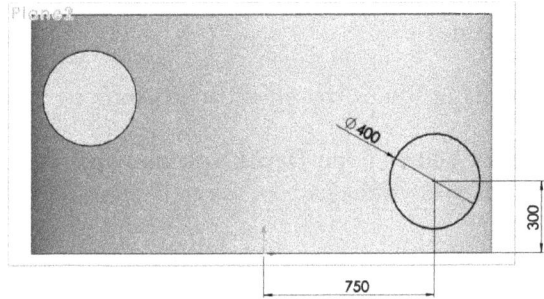

Figure 3-35 *Sketch created on plane*

6. Change the current view to the isometric view and invoke the **Extruded Boss/Base** tool.

7. Right-click in the drawing area and choose the **Up To Surface** option from the shortcut menu; you are prompted to select a face or a surface to specify the first direction. Also, the **Face/Plane** selection box is displayed below the **Direction of Extrusion** selection box in the **Direction 1** rollout.

8. Select the upper curved surface of the model using the left mouse button. You will observe in the preview that the feature is extruded up to the selected surface.

9. Choose the **OK** button from the **Boss-Extrude PropertyManager**, as shown in Figure 3-36.

    For better display of the model turn off the plane display.

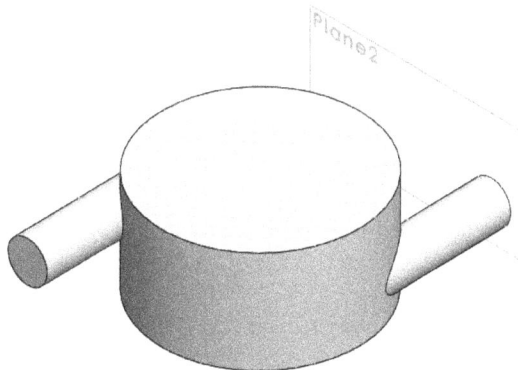

Figure 3-36 *Inlet Pipe 2 created*

Creating the Inlet Pipe 3

Now, you need to create a cylindrical feature to get the required shape of the base feature. The sketch for this feature is created on the bottom face of the base feature.

1. Select the bottom face of the base feature and invoke the sketching environment. Set the current view normal to the eye view.

2. Draw the sketch of the circle and apply required relations to the sketch, as shown in Figure 3-37.

3. Change the current view to the isometric view and invoke the **Extruded Boss/Base** tool.

4. Set **1000** in the **Depth** spinner and choose the **OK** button from the **Boss-Extrude PropertyManager**; the selected contour is extruded, refer to Figure 3-38.

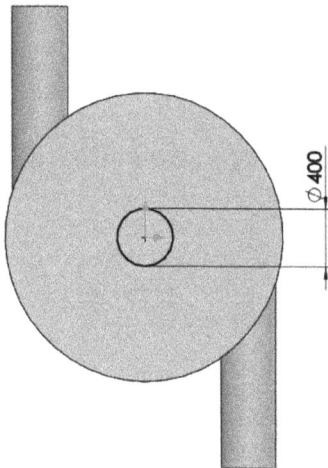

Figure 3-37 Sketch created on bottom face   *Figure 3-38 Inlet Pipe 3 created*

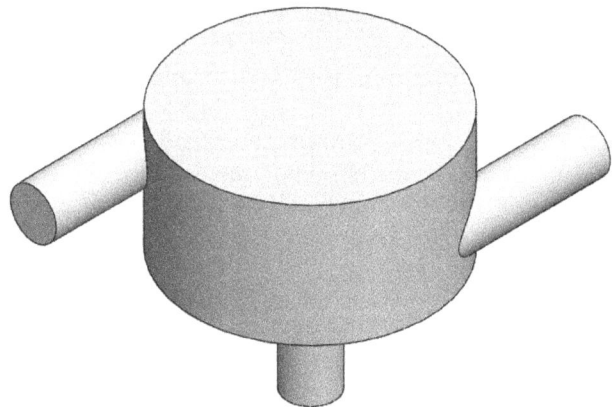

Creating the Shell Feature for Fluid Flow

The **Shell** tool is used to scoop out material from the model leaving behind a thin-walled hollow part. This hollow space is used for fluid flow.

1. Choose the **Shell** button from the **Features CommandManager**; the **Shell PropertyManager** is displayed and you are prompted to select the faces to be removed.

2. Rotate the model and select the faces to be removed, as shown in Figure 3-39; the names of the selected faces are displayed in the **Faces to Remove** selection box.

3. Set the value to **10** in the **Thickness** spinner and choose the **OK** button from the **Shell PropertyManager**. The model after creating the shell feature is shown in Figure 3-40.

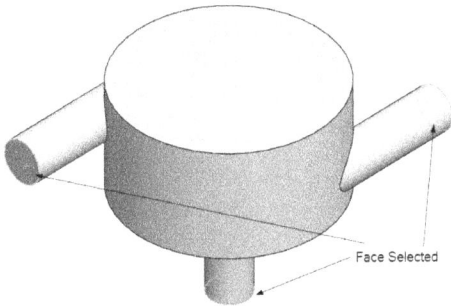

Figure 3-39 Face to be shell   *Figure 3-40* Model after shell

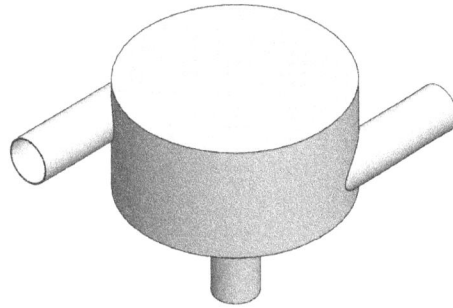

Creating the Lids by Using the Extruded Feature

Next, you need to create the lid at the inlet and outlet of the model.

1. Invoke the **Extruded Boss/Base** tool, select the inlet pipe 1 face as the sketching plane, and create the sketch using the standard sketching tools. The sketch consists of a circle having 380 mm diameter.

2. Exit the sketching environment; the **Boss-Extrude PropertyManager** is displayed.

3. Set **50** in the **Depth** spinner and clear the **Merge Result** check box. Choose the **OK** button from the **Boss-Extrude PropertyManager**; the selected contour is extruded, refer to Figure 3-41.

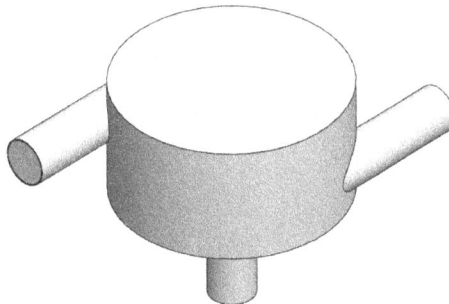

Figure 3-41 Lid added to the model

4. Similarly, create Lid 2 and Lid 3 for the inlet 2 and outlet 1.

    Next, rename the tree items to Lid 1, Lid 2, and Lid 3.

5. Right-click on **Boss-Extrude5** and choose the **Rename tree item** option. Type the name **Lid 1**; the name is updated in the design tree. Refer to Figure 3-42 for design tree.

    Similarly, you can rename the other extruded feature as Lid 2 and Lid 3.

Saving the Model

1. Save the part document with the name **c03_tut01** at the following location: *SOLIDWORKS_Flow\resources\c03*.

2. Choose **File > Close** from the SOLIDWORKS menus to close the document.

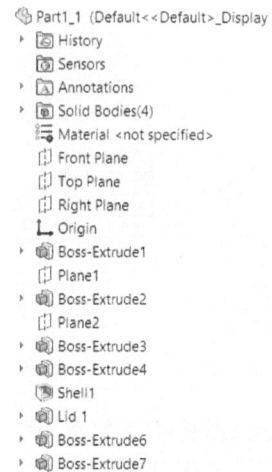

Figure 3-42 *Feature Manager Design Tree*

Tutorial 2

In this tutorial, you will create a model and make it ready for flow simulation. The model and its drawing are shown in Figure 3-43 and 3-44.   **(Expected time: 30 min)**

    The following steps are required to complete this tutorial:

a. Start SOLIDWORKS and then start a new part document.
b. Create the base feature.
c. Create the pipe channel.
d. Connect the base feature and pipe channel for heat transfer.
e. Create Lid 1.
f. Create Lid 2.

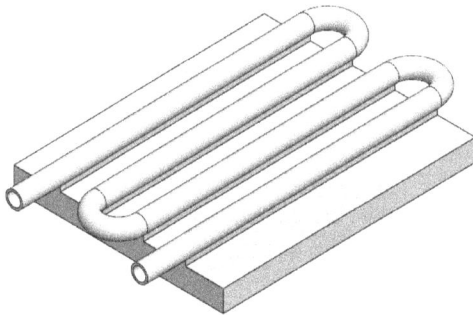

Figure 3-43 *Solid model for Tutorial 2*

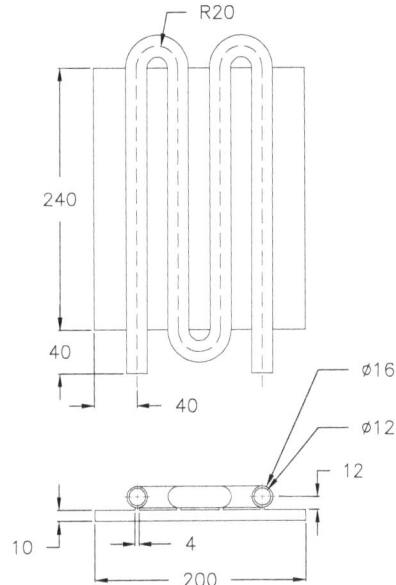

Figure 3-44 *Sketch for Tutorial 2*

Starting SOLIDWORKS and then a New SOLIDWORKS Document

1. Start SOLIDWORKS by double-clicking on the shortcut icon of SOLIDWORKS 2023 on the desktop of your computer; the SOLIDWORKS 2023 window along with the **Welcome - SOLIDWORKS 2023** dialog box is displayed.

2. Choose the **Part** button available in the **New** area of the **Home** tab in the **Welcome - SOLIDWORKS 2023** dialog box; a new **SOLIDWORKS** part document starts.

    Next, you need to invoke the sketching environment.

3. Choose the **Sketch** tab from the **CommandManager** if not selected by default. Next, choose the **Sketch** button from the **Sketch CommandManager**; the **Edit Sketch PropertyManager** is invoked and you are prompted to select the plane to create the sketch.

4. Select the **Top Plane** from the drawing area; the sketching environment is invoked and the plane gets oriented normal to the view. You will notice that red colored arrows are displayed at the center of the screen indicating that you are in the sketching environment. Also, the confirmation corner with the Exit Sketch and Cancel options is displayed on the upper right corner in the graphics area. The screen display in the sketching environment of SOLIDWORKS 2023 is shown in Figure 3-45.

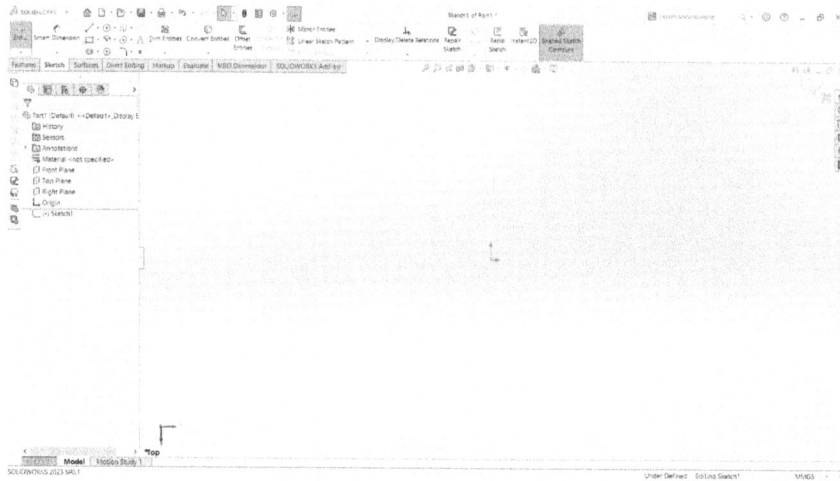

Figure 3-45 Screen display in the sketching environment

Drawing the Sketch

Next, you need to draw the sketch using the sketch tools. As it is evident from Figure 3-44, the sketch needs to be drawn using the **Rectangle** tool to create the sketch of base feature.

1. Choose the **Center Rectangle** tool from the **Rectangle** flyout of the **Sketch CommandManager** to draw a rectangle by specifying the center and one of the corners; the **Rectangle PropertyManager** is displayed on the left of the drawing area. Also, the arrow cursor changes to the rectangle cursor.

2. Move the rectangle cursor to the origin and click when an orange circle is displayed to specify the center of the rectangle.

3. Move the cursor toward the right and click when a dimension is displayed above the rectangle cursor close to the dimension of the base feature.

   Next, add dimension to the sketch.

4. Choose the **Smart Dimension** tool and select the vertical line; the dimension is attached to the cursor. Place the dimension; the **Modify** dialog box is displayed with the default value. Enter **240** in the **Modify** dialog box and press ENTER.

5. Next, select the horizontal line; the dimension is attached to the cursor. Place the dimension; the **Modify** dialog box is displayed with the default value. Enter **200** in the **Modify** dialog box and press ENTER. Refer to Figure 3-46 for the sketch.

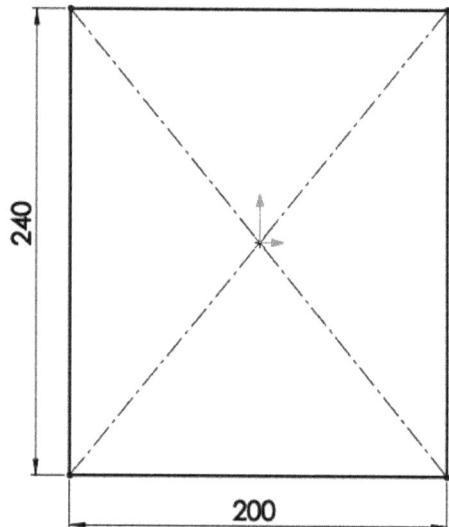

Figure 3-46 Sketch after applying dimension

Extruding the Sketch

The next step after creating the sketch is to extrude it. You can extrude the sketch using the **Extruded Boss/Base** tool.

1. Choose the **Features** tab from the CommandManager to display the **Features CommandManager**. Then, choose the **Extruded Boss/Base** button; the sketch is automatically oriented to the trimetric view and the **Boss-Extrude PropertyManager** is displayed.

   As you are converting the closed sketch into a feature, only the **Direction 1** rollout is displayed in the **Boss-Extrude PropertyManager**. Also, a preview of the feature is displayed in the temporary shaded graphics with the default values.

2. Set **10** in the **Depth** spinner as the depth in the **Direction 1** rollout and then choose the **OK** button to extrude the sketch. Refer to Figure 3-47 for the model.

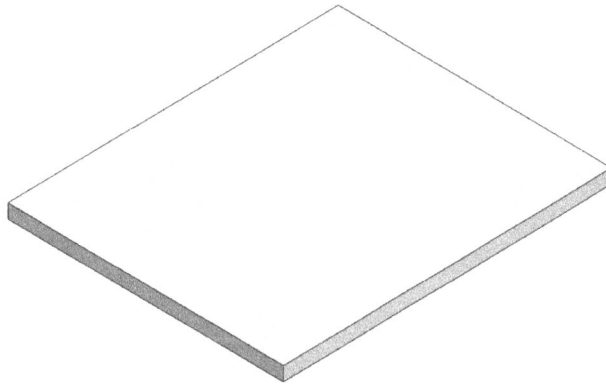

Figure 3-47 Model created by extruding the sketch

Creating the Plane

Next, create a plane at an offset of 6mm from the top face of the base feature.

1. Choose the **Plane** tool from the **Reference Geometry** flyout in the **Features CommandManager**; the **Plane PropertyManager** is displayed.

2. Select the top face of the base feature; the selected face is displayed in the **First Reference** selection box.

3. Enter **12** in the **Offset distance** spinner to create the plane at an offset distance. Choose the **OK** button to close the **Plane PropertyManager**, refer to Figure 3-48 for the created plane.

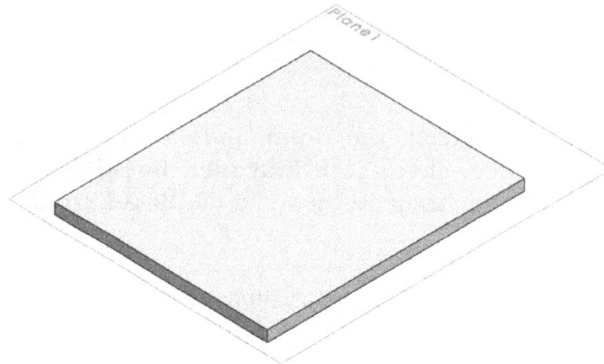

Figure 3-48 *Creating a plane offset from face*

Creating the Channel

Next, you need to draw the the sketch for channel using the sketch tools. As it is evident from Figure 3-44, the sketch needs to be drawn using the **Arc** and **Line** tools. After creating the sketch, you need to sweep the circular profile along that sketch.

1. Choose the **Sketch** button from the **Sketch CommandManager** and select Plane 1 to invoke the sketching environment.

2. Press and hold the right-mouse button and drag the cursor to the right; the mouse gesture is displayed with sketching tools. Move the cursor over the **Line** tool; the **Line** tool is invoked and the **Insert Line PropertyManager** is displayed.

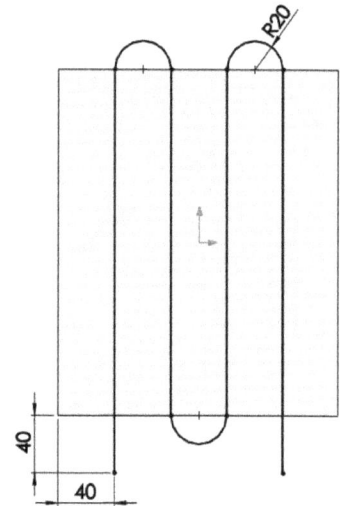

3. Create the sketch of the channel of the model on the offset plane created earlier. Apply the required relations and dimensions to the sketch, as shown in Figure 3-49. Make sure that you do not exit the sketching environment.

Figure 3-49 *The sketch on a plane*

Next, the sweep feature that you need to create is a thin circular sweep feature. You need to use the **Thin Feature** rollout to specify the parameters of the sweep feature.

4. Choose the **Swept Boss/Base** tool from the **Features CommandManager**; the **Sweep PropertyManager** is displayed.

5. Select the **Circular Profile** radio button from the **Profile and Path** rollout. As a result, the **Path** selection box and the **Diameter** spinner is displayed.

6. Select the sketch from the drawing area; the name of the sketch is displayed in the **Path** selection box. Also, set the value of diameter in the **Diameter** spinner to **16**.

7. Select the check box on the left of the **Thin Feature** rollout; the **Thin Feature** rollout is invoked.

8. Set the value of thickness in the **Thickness** spinner to **2**.

9. Choose the **OK** button from the **Sweep PropertyManager** or choose **OK** from the confirmation corner. Refer to Figure 3-50 for the model.

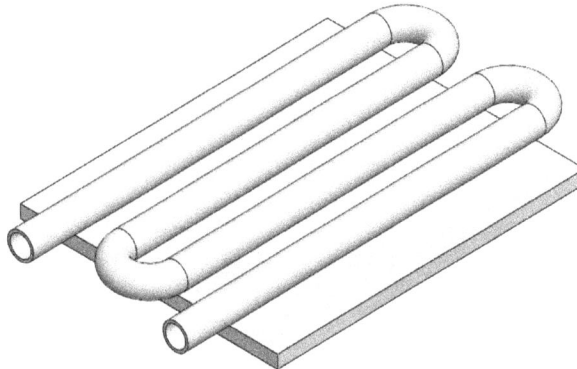

Figure 3-50 Model created by sweeping

Connecting Plate and Pipe Channel Through Joint

Next, you need to connect the plate and pipe channel so that heat transfer takes place between these two.

1. Choose the **Sketch** button from the **Sketch CommandManager**; the **Edit Sketch PropertyManager** is invoked and you are prompted to select the plane to create the sketch.

2. Select the front face of the plate from the drawing area; the sketching environment is invoked and the plane gets oriented normal to the view.

3. Create the sketch of joint and dimension it as shown in Figure 3-51. Similarly, create the sketch of other joints and dimension them.

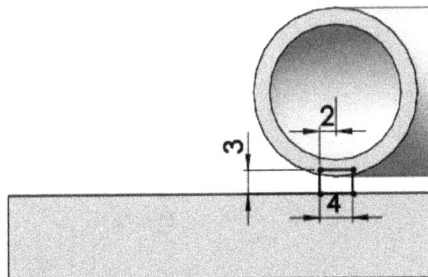

Figure 3-51 Creating the sketch of joint

4. Choose the **Exit Sketch** button to exit the sketching environment.

   Next, you need to extrude the sketch upto the back face of the plate.

5. Invoke the **Extruded Boss/Base** tool and select the sketch if it is not selected.

6. Select the **Up To Surface** option from the **End Condition** drop-down in the **Direction 1** rollout. Select the back face of the plate, a preview of the feature is displayed.

7. Choose the **OK** button to extrude the sketch. Refer to Figure 3-52 for the joint created by extruding the sketch.

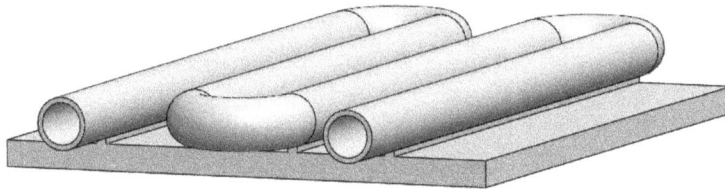

Figure 3-52 Joint created by extrusion

Creating the Lids by Using the Extruded Feature

Next, you need to create Lid 1 and Lid 2 which helps you to specify the inlet and outlet conditions in flow simulation environment.

1. Invoke the **Extruded Boss/Base** tool, select the inlet pipe 1 face as the sketching plane, and create the sketch using the standard sketching tools. The sketch consists of a circle having 12 mm diameter.

2. Exit the sketching environment; the **Boss-Extrude PropertyManager** is displayed.

3. Set **5** in the **Depth** spinner and clear the **Merge Result** check box. Choose the **OK** button from the **Boss-Extrude PropertyManager**; the selected contour is extruded, refer to Figure 3-53.

4. Similarly, create Lid 2 for outlet of pipe channel, refer to Figure 3-54.

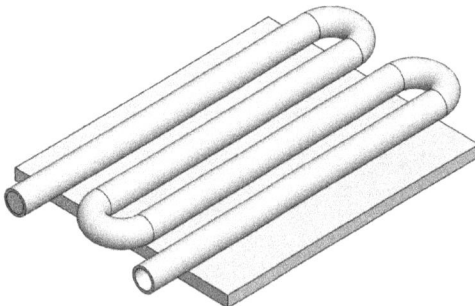

Figure 3-53 Creating Lid 1 at inlet

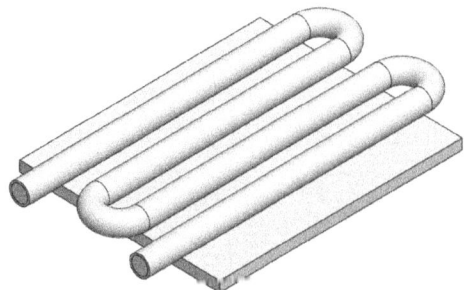

Figure 3-54 Creating Lid 2 at outlet

Next, rename the tree items to Lid 1, and Lid 2.

5.   Right-click on **Boss-Extrude5** and select the **Rename tree item** option. Type the name **Lid 1**; the name is updated in the tree.

Similarly, you can rename the other extruded feature as Lid 2.

Saving the Model

1.   Save the part document with the name **c03_tut02** at the following location: *SOLIDWORKS_ Flow\resources\c03*.

2.   Choose **File > Close** from the SOLIDWORKS menus to close the document.

Tutorial 3

In this tutorial, you will create a Ahmed body model. The model and its drawing are shown in Figure 3-55 and 3-56.   **(Expected time: 30 min)**

The following steps are required to complete this tutorial:

a.   Start SOLIDWORKS and then start a new part document.
b.   Create the base feature.
c.   Apply fillet to the feature.
d.   Create the cut feature.
e.   Create legs of body by extrusion.

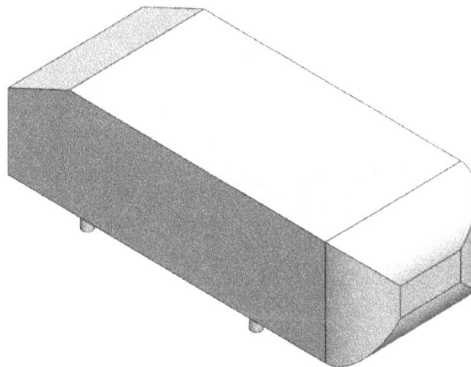

Figure 3-55 Solid model for Tutorial 3

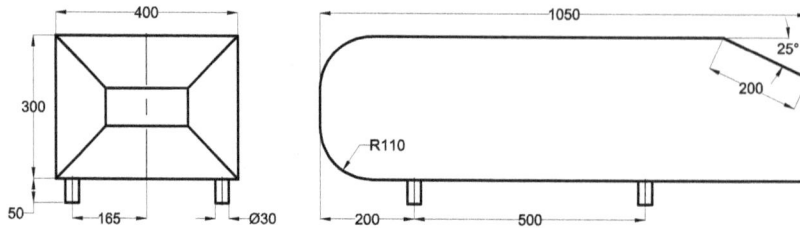

Figure 3-56 *Sketch for Tutorial 3*

Starting SOLIDWORKS and then a New SOLIDWORKS Document

1. Start SOLIDWORKS by double-clicking on the shortcut icon of SOLIDWORKS 2023 on the desktop of your computer; the SOLIDWORKS 2023 window along with the **Welcome - SOLIDWORKS 2023** dialog box is displayed.

2. Choose the **Part** button available in the **New** area of the **Home** tab in the **Welcome - SOLIDWORKS 2023** dialog box; a new SOLIDWORKS part document starts.

   Next, you need to invoke the sketching environment.

3. Choose the **Sketch** tab from the **CommandManager** if not selected by default. Next, choose the **Sketch** button from the **Sketch CommandManager**; the **Edit Sketch PropertyManager** is invoked and you are prompted to select the plane to create the sketch.

4. Select the Right Plane from the drawing area; the sketching environment is invoked and the plane gets oriented normal to the view. You will notice that the red colored arrows are displayed at the center of the screen indicating that you are in the sketching environment. Also, the confirmation corner with the Exit Sketch and Cancel options is displayed on the upper right corner in the graphics area. The screen display in the sketching environment of SOLIDWORKS 2023 is shown in Figure 3-57.

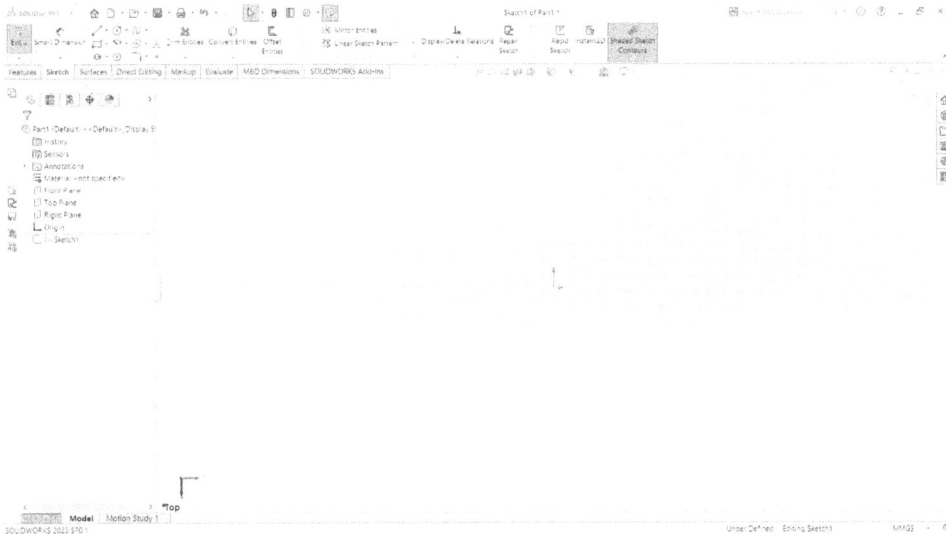

Figure 3-57 *Screen display in the sketching environment*

Drawing the Sketch

Next, you need to draw the sketch using the sketch tools. As it is evident from Figure 3-56, the sketch needs to be drawn using the **Rectangle** tool to create the sketch of base feature.

1. Choose the **Corner Rectangle** tool from the **Rectangle** flyout of the **Sketch CommandManager** to draw a rectangle by specifying the center and one of the corners; the **Rectangle PropertyManager** is displayed on the left of the drawing area. Also, the arrow cursor changes to the rectangle cursor.

2. Create the rectangle shown in Figure 3-58.

   Next, add dimension to the sketch.

3. Choose the **Smart Dimension** tool and add a dimension to the sketch, refer to Figure 3-58 for the sketch.

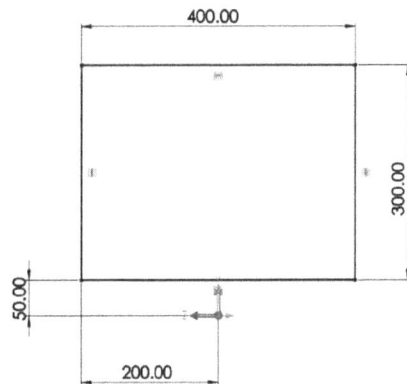

Figure 3-58 *Sketch after applying dimension*

Extruding the Sketch

The next step after creating the sketch is to extrude it. You can extrude the sketch using the **Extruded Boss/Base** tool.

1. Choose the **Features** tab from the CommandManager to display the **Features CommandManager**. Then, choose the **Extruded Boss/Base** button; the sketch is automatically oriented to the trimetric view and the **Boss-Extrude PropertyManager** is displayed.

As you are converting the closed sketch into a feature, only the **Direction 1** rollout is displayed in the **Boss-Extrude PropertyManager**. Also, a preview of the feature is displayed in the temporary shaded graphics with the default values.

2. Set **1050** in the **Depth** spinner as the depth in the **Direction 1** rollout and then choose the **OK** button to extrude the sketch. Refer to Figure 3-59 for the model.

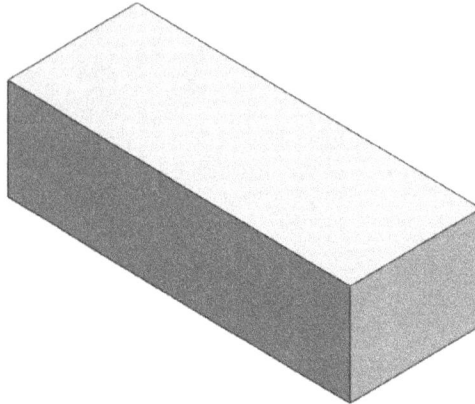

Figure 3-59 Model created by extruding the sketch

Creating the Fillet Features

After creating the base feature, you need to add fillets to the model. In this model, you need to add four fillet features.

1. Choose the **Fillet** button from the **Features CommandManager** to invoke the **Fillet PropertyManager**.

2. Select the edges of the model, as shown in Figure 3-60.

    On selecting the edges of the model, the preview of the fillet with default values and the radius callout is displayed in the drawing area.

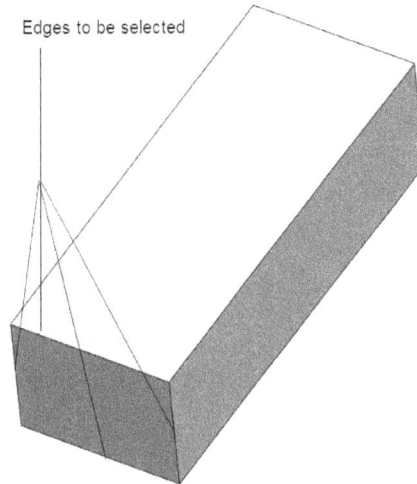

Figure 3-60 Edges to be selected

3. Set the value in the **Radius** spinner to **110** and choose the **OK** button from the **Fillet PropertyManager**. Figure 3-61 shows the model after adding the fillet feature.

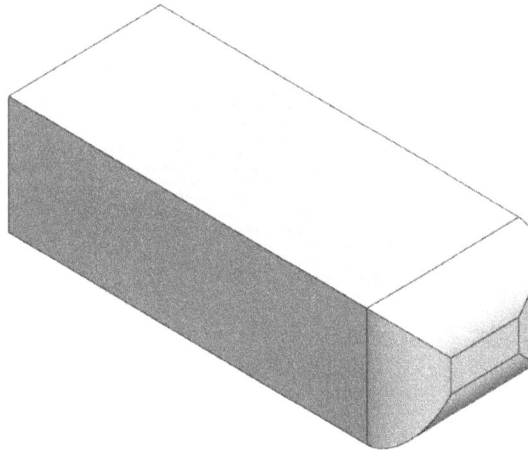

Figure 3-61 Fillet added to the model

Creating the Cut Feature

The next feature that you will create is a cut feature.

1. Select the Front plane from the drawing area as the sketching plane and press **S** on the keyboard. Invoke the **Extruded Cut** tool from the Shortcut toolbar.

2. Draw a sketch using the standard sketch tools, refer to Figure 3-62.

Figure 3-62 *Sketch for the cut feature*

3. Exit the sketching environment and choose the **Through All** option from the **End Condition** drop-down list of the **Direction 1** and **Direction 2** rollouts.

4. Choose the **OK** button from the **Cut-Extrude1 PropertyManager,** refer to Figure 3-63 for the model.

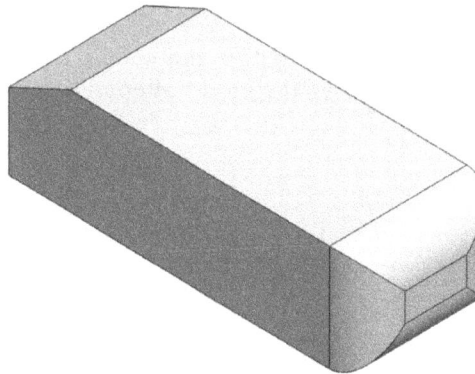

Figure 3-63 *Model after the cut feature*

Creating the Extruded Feature

Next, you need to create legs of the body using the **Extruded Boss/Base** tool.

1. Create the sketch on the bottom face of the body, refer to Figure 3-64.

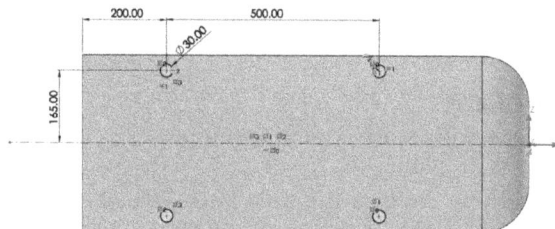

Figure 3-64 *Sketch for the extruded feature*

2. Choose the **Features** tab from the CommandManager to display the **Features CommandManager**. Then, choose the **Extruded Boss/Base** button; the sketch is automatically oriented to the trimetric view and the **Boss-Extrude PropertyManager** is displayed.

3. Set **50** in the **Depth** spinner as the depth in the **Direction 1** rollout and then choose the **OK** button to extrude the sketch. Refer to Figure 3-65 for the model.

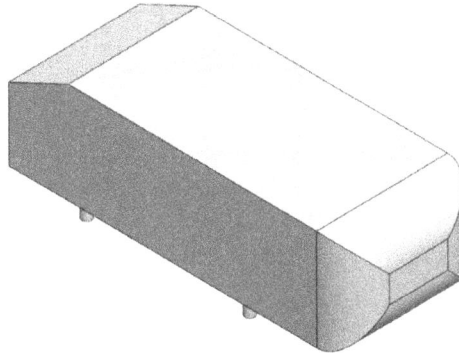

Figure 3-65 Creating legs at the bottom of the body

Saving the Model

1. Save the part document with the name **c03_tut03** at the following location: *\SOLIDWORKS_ Flow\resources\c03*.

2. Choose **File > Close** from the SOLIDWORKS menus to close the document.

Tutorial 4

In this tutorial, you will create a centrifugal pump model which is a multi body parts. The model and its drawing are shown in Figure 3-66 and 3-67.   **(Expected time: 30 min)**

The following steps are required to complete this tutorial:

a. Start SOLIDWORKS and then start a new part document.
b. Create body 1.
c. Create body 2.
d. Create the inlet and outlet for body 2.
e. Create the shell feature.

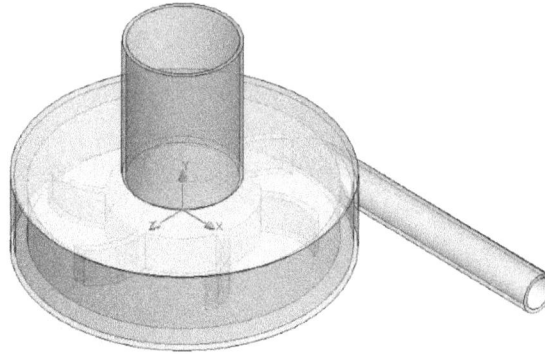

Figure 3-66 *Solid model for Tutorial 4*

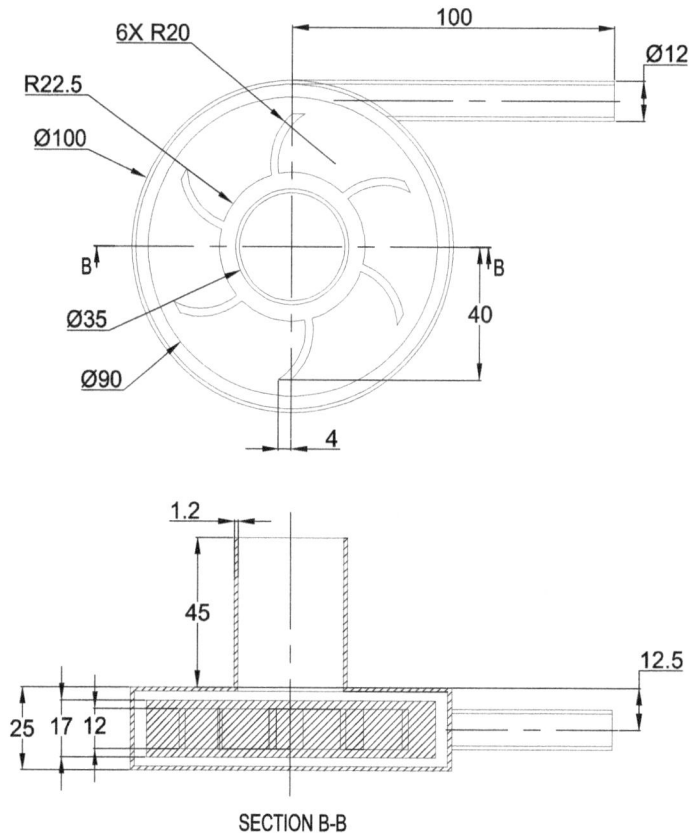

Figure 3-67 *Sketch for Tutorial 4*

Starting SOLIDWORKS and then a New SOLIDWORKS Document

1. Start SOLIDWORKS by double-clicking on the shortcut icon of SOLIDWORKS 2023 on the desktop of your computer; the SOLIDWORKS 2023 window along with the **Welcome - SOLIDWORKS 2023** dialog box is displayed.

2. Choose the **Part** button available in the **New** area of the **Home** tab in the **Welcome - SOLIDWORKS 2023** dialog box; a new **SOLIDWORKS** part document starts.

    Next, you need to invoke the sketching environment.

3. Choose the **Sketch** tab from the **CommandManager** if not selected by default. Next, choose the **Sketch** button from the **Sketch CommandManager**; the **Edit Sketch PropertyManager** is invoked and you are prompted to select the plane to create the sketch.

4. Select the **Right Plane** from the drawing area; the sketching environment is invoked and the plane gets oriented normal to the view. You will notice that the red colored arrows are displayed at the center of the screen indicating that you are in the sketching environment. Also, the confirmation corner with the Exit Sketch and Cancel options is displayed on the upper right corner in the graphics area. The screen display in the sketching environment of SOLIDWORKS 2023 is shown in Figure 3-68.

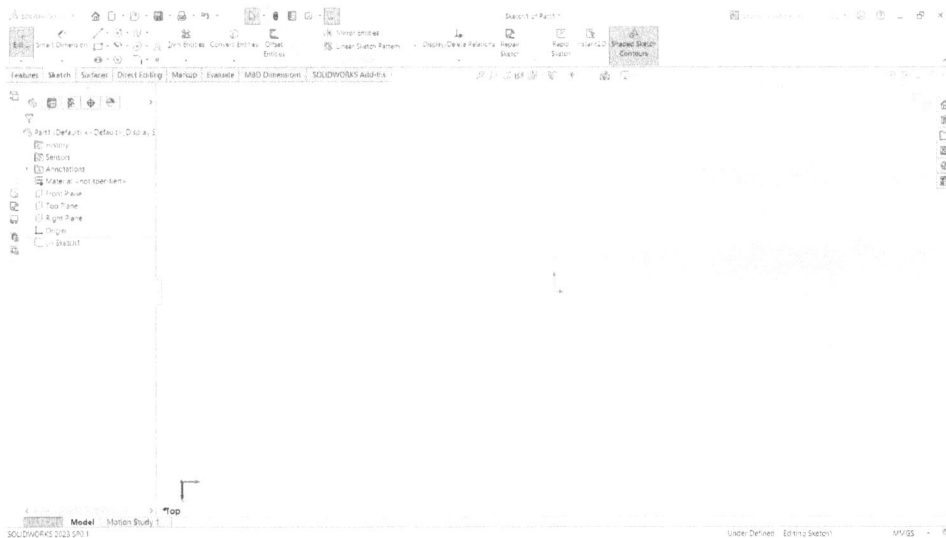

Figure 3-68 *Screen display in the sketching environment*

Drawing the Sketch

Next, you need to draw the sketch using the sketch tools. As it is evident from Figure 3-56, the sketch needs to be drawn using the **Rectangle** tool to create the sketch of base feature.

1. Create the sketch using the line, circle and arc tools, as shown in Figure 3-69.

    Next, add dimension to the sketch.

2. Choose the **Smart Dimension** tool and add a dimension to the sketch, refer to Figure 3-69 for the sketch.

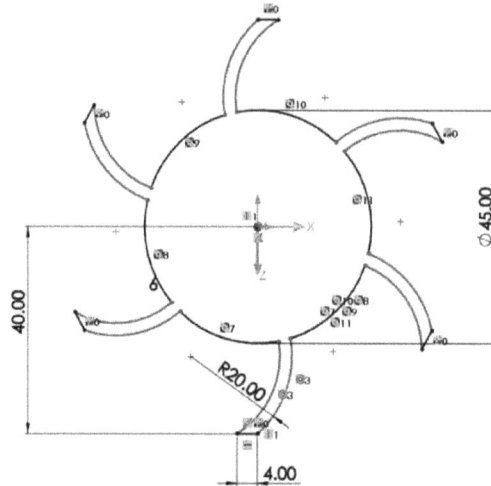

Figure 3-69 *Sketch after applying dimension*

Extruding the Sketch

The next step after creating the sketch is to extrude it. You can extrude the sketch using the **Extruded Boss/Base** tool.

1. Choose the **Features** tab from the CommandManager to display the **Features CommandManager**. Then, choose the **Extruded Boss/Base** button; the sketch is automatically oriented to the trimetric view and the **Boss-Extrude PropertyManager** is displayed.

    As you are converting the closed sketch into a feature, only the **Direction 1** rollout is displayed in the **Boss-Extrude PropertyManager**. Also, a preview of the feature is displayed in the temporary shaded graphics with the default values.

2. Set **1050** in the **Depth** spinner as the depth in the **Direction 1** rollout and then choose the **OK** button to extrude the sketch. Refer to Figure 3-70 for the model.

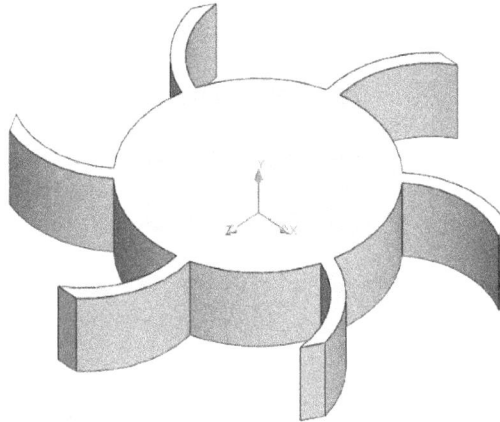

Figure 3-70 *Model created by extruding the sketch*

Drawing the Casing Sketch

Next, you need to draw the sketch using the sketch tools.

1. Create the sketch using the **Circle** tool as shown in Figure 3-71.

   Next, add dimension to the sketch.

2. Choose the **Smart Dimension** tool and add a dimension to the sketch, refer to Figure 3-71 for the sketch.

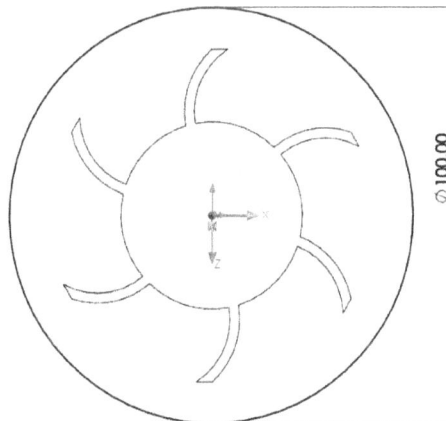

Figure 3-71 *Sketch after applying dimension*

Extruding the Casing Sketch

The next step after creating the sketch is to extrude it. You can extrude the sketch using the **Extruded Boss/Base** tool.

1. Choose the **Features** tab from the CommandManager to display the **Features CommandManager**. Then, choose the **Extruded Boss/Base** button; the sketch is automatically oriented to the trimetric view and the **Boss-Extrude PropertyManager** is displayed.

   As you are converting the closed sketch into a feature, only the **Direction 1** rollout is displayed in the **Boss-Extrude PropertyManager**. Also, a preview of the feature is displayed in the temporary shaded graphics with the default values.

2. Set **25** in the **Depth** spinner as the depth and select the **Midplane** option in the **End Condition** drop-down list in the **Direction 1** rollout and then choose the **OK** button to extrude the sketch. Refer to Figure 3-72 for the model.

   You need to ensure that the **Merge result** check box is not selected and adjust the transparency of the model in order to see inside it.

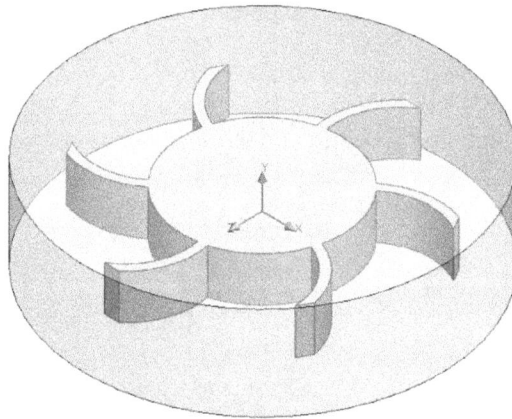

Figure 3-72 Model created by extruding the sketch

Drawing the Inlet Sketch

Next, you need to draw the sketch on top face of the casing using the sketch tools.

1. Create the sketch using the **Circle** tool, as shown in Figure 3-73.

   Next, add dimension to the sketch.

2. Choose the **Smart Dimension** tool and add a dimension to the sketch, refer to Figure 3-73 for the sketch.

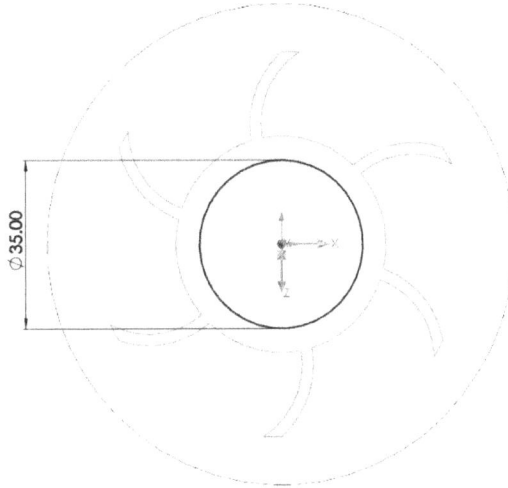

Figure 3-73 Sketch after applying dimension

Creating the Inlet Extruded Feature

Next, you need to create legs of the body using the **Extruded Boss/Base** tool.

1. Choose the **Features** tab from the CommandManager to display the **Features CommandManager**. Then, choose the **Extruded Boss/Base** button; the sketch is automatically oriented to the trimetric view and the **Boss-Extrude PropertyManager** is displayed.

2. Set **45** in the **Depth** spinner as the depth in the **Direction 1** rollout and then choose the **OK** button to extrude the sketch. Refer to Figure 3-74 for the model. Ensure that the **Merge result** check box is selected.

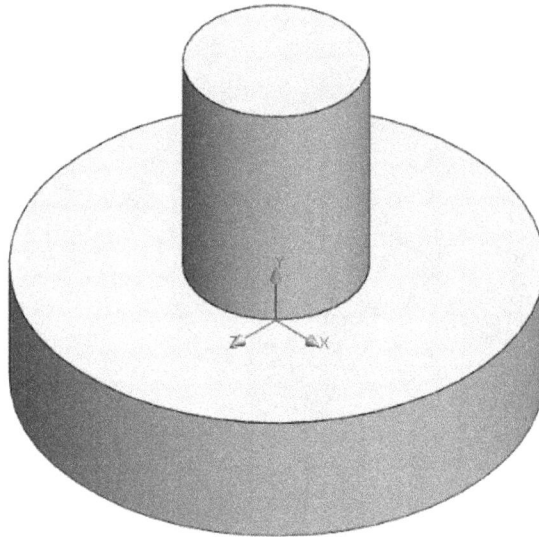

Figure 3-74 *Model created by extruding the sketch*

Drawing the Outlet Sketch

Next, you need to draw the sketch using the sketch tools. Before creating the sketch, you need to create a plane 100 mm offset from the Right plane.

1. Choose the **Plane** tool from the **Reference Geometry** drop-down; the **Plane PropertyManager** is displayed.

2. Select the Right Plane from the drawing area; the name of the plane will be selected in the **First Reference** edit box. Enter **100** in the **Offset distance** edit box. Refer to Figure 3-75 for the plane.

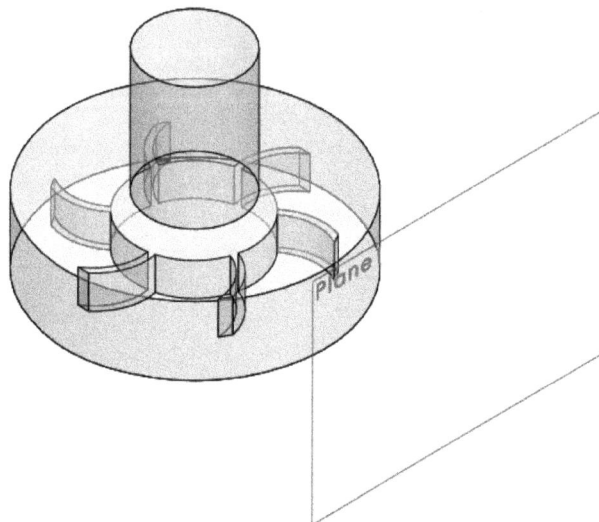

Figure 3-75 *The plane created*

Next, create the sketch on the plane.

3. Create the sketch using the **Circle** tool, as shown in Figure 3-76.

Next, add dimension to the sketch.

4. Choose the **Smart Dimension** tool and add a dimension to the sketch, refer to Figure 3-76 for the sketch.

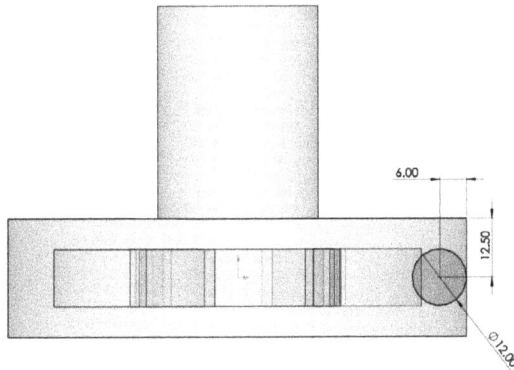

Figure 3-76 *Sketch after applying dimensions*

Creating the Outlet Extruded Feature

Next, you need to create legs of the body using the **Extruded Boss/Base** tool.

1. Choose the **Features** tab from the CommandManager to display the **Features CommandManager**. Then, choose the **Extruded Boss/Base** button; the sketch is automatically oriented to the trimetric view and the **Boss-Extrude PropertyManager** is displayed.

2. Select the **Up To Body** option from the **End Condition** drop-down in the **Direction 1** rollout and then select the casing from the drawing area which will be selected in Solid/Surface Body area in the **Direction 1** rollout. Choose the **OK** button to extrude the sketch. Refer to Figure 3-77 for the model. Ensure that the **Merge result** check box is selected.

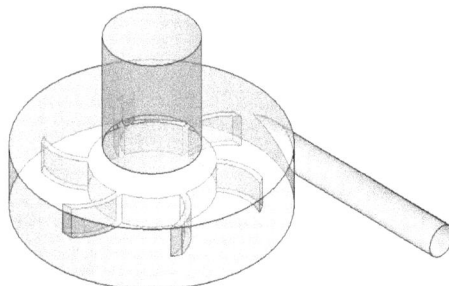

Figure 3-77 *Model created by extruding the sketch*

Creating the Shell Feature

It is evident from Figure 3-66 that a shell feature is required to create a thin-walled structure.

1. Choose the **Shell** button from the **Features CommandManager**; the **Shell PropertyManager** is displayed and you are prompted to select the faces to be removed.

2. Select the inlet and outlet faces of the model; the selected face is displayed in the **Faces to Remove** selection box, refer to Figure 3-78.

3. Set the value to **1.2** mm in the **Thickness** spinner of the PropertyManager. Choose the **OK** button from the **Shell PropertyManager**.

The model after creating the shell feature is shown in Figure 3-79.

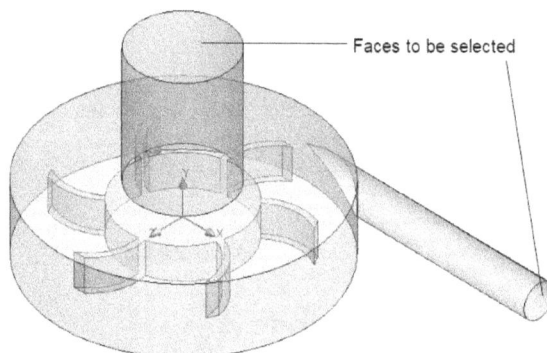

Figure 3-78 Faces selected to be removed

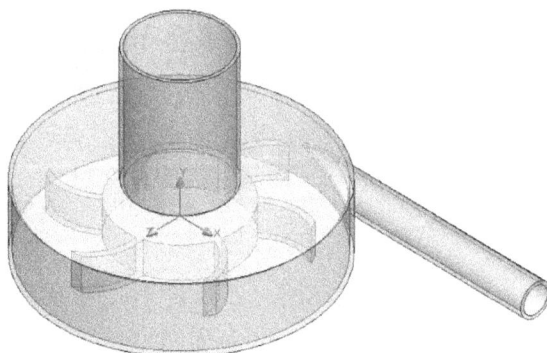

Figure 3-79 Model after creating the shell feature

Creating the Rotating Region

Next, you need to create the rotating region by creating and extruding the sketch.

1. Create the sketch on the Top Plane, as shown in Figure 3-80.

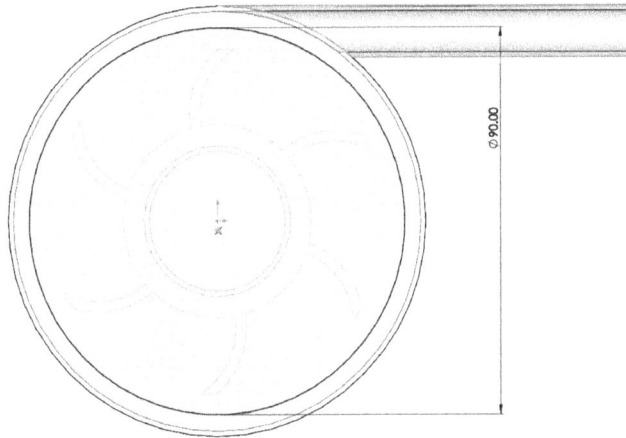

Figure 3-80 *Sketch after applying dimensions*

2. Create the solid model whose extrusion depth is 17 mm, symmetrically, as shown in Figure 3-81.

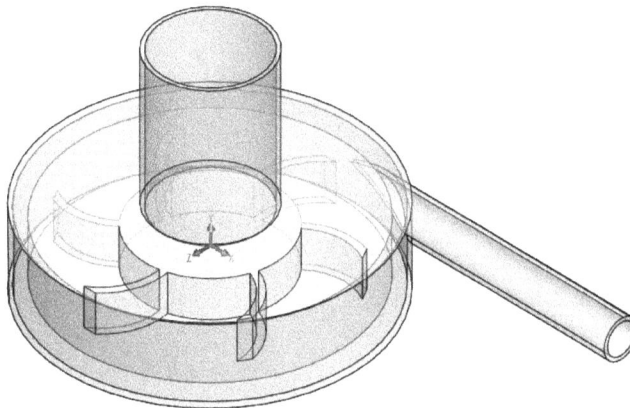

Figure 3-81 *The Solid model of rotating region*

After making the rotating region, close the inlet and outlet openings with a lid, just as in tutorial 1 and tutorial 2.

Saving the Model

1. Save the part document with the name **c03_tut04** at the following location: *\SOLIDWORKS_ Flow\resources\c03*.

2. Choose **File > Close** from the SOLIDWORKS menus to close the document.

Self-Evaluation Test

Answer the following questions and then compare them to those given at the end of this chapter:

1. The _____ tool is used to extrude a sketch in SOLIDWORKS.

2. The _____ tool is used to revolve a sketch in SOLIDWORKS.

3. The _____ option is used to create the base feature by extruding the sketch equally in both the directions of the plane on which the sketch is drawn.

4. The _____ button is used to specify a draft angle while extruding a sketch.

5. The Extruded Boss/Base tool is also used to extrude the open profile. (T/F)

6. The thin extruded features can be created using a closed sketch. (T/F)

Review Questions

Answer the following questions:

1. Which of the following PropertyManagers is displayed when you choose the **Extruded Boss/Base** tool from the **Features CommandManager**?

   (a) **Boss**   (b) **Extrude**
   (c) **Boss-Extrude**   (d) None of these

2. Which of the following PropertyManagers is displayed on the left of the drawing area when you choose the **Swept Boss/Base** tool from the **Features CommandManager**?

   (a) **Swept**   (b) **Boss**
   (c) **Sweep**   (d) None of these

3. Which of the following options is used to create the extrude thickness?

   (a) **Thin Feature**   (b) **Thin**
   (c) **Shell**   (d) None of these

4. You cannot scoop out material from a model using the **Shell** tool. (T/F)

5. You can create shell features with different thickness of different faces. (T/F)

EXERCISES

Exercise 1

Create the solid model of the sphere shown in Figure 3-82. The sketch of the sphere is shown in Figure 3-83.   **(Expected time: 30 min)**

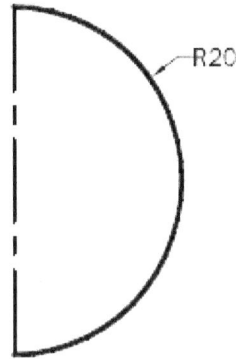

Figure 3-82 Solid model for Exercise 1   *Figure 3-83 Sketch for Exercise 1*

Exercise 2

Create the solid model of the cylinder shown in Figure 3-84. The sketch of the cylinder is shown in Figure 3-85.   **(Expected time: 10 min)**

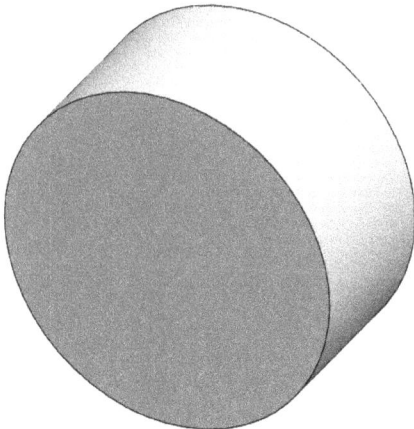

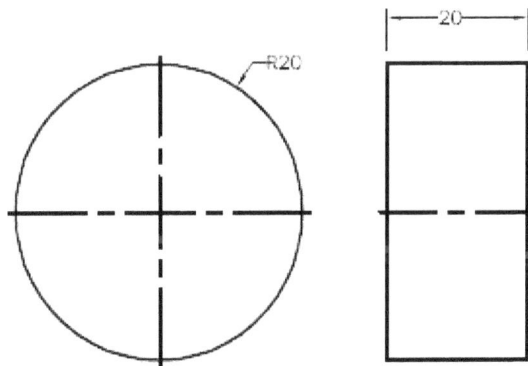

Figure 3-84 Solid model for Exercise 2   *Figure 3-85 Sketch for Exercise 2*

Answers to Self-Evaluation Test

1. Extruded Boss/Base, 2. Revolved Boss/Base, 3. Mid Plane, 4. Draft On/Off, 5. T, 6. T

Chapter 4

Creating a Flow Simulation Project

Learning Objectives

After completing this chapter, you will be able to:

- *Specify the unit system*
- *Specify the analysis type*
- *Specify a fluid for a project*
- *Define a wall type and initial conditions*

CREATING FLOW SIMULATION PROJECT

In this chapter, you will learn to create a flow simulation project. To initiate a flow simulation project, you must utilize either the **Wizard**, **New Project**, or **Clone Project** tool. This chapter covers the **Wizard** tool, which will be discussed next.

WIZARD

CommandManager:	Flow Simulation > Wizard
SOLIDWORKS Menus:	Tools > Flow Simulation > Project > Wizard
Toolbar:	Flow Simulation Main > Wizard

The **Wizard** tool from the **Flow Simulation CommandManager** is used to create a project. To create a project, choose the **Flow Simulation** tab from the **CommandManager** to display the **Flow Simulation CommandManager**. Next, choose the **Wizard** tool from the **Flow Simulation CommandManager**; the **Wizard - Project Name** page is displayed, refer to Figure 4-1. The options of this page are discussed next.

In this page, you can specify the configuration and name for the project. There are the **Project** and **Configuration to add the project** areas in this page and they are discussed next.

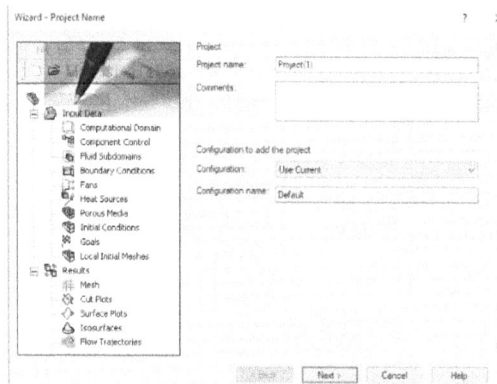

Figure 4-1   The Wizard - Project Name page

Project

In this area, you can specify a name for the project in the **Project name** edit box and also write the comments about the project in the **Comments** edit box.

Configuration to add the project

This area is used to specify the required configuration with the geometry for an analysis. You can select one of the options, **Use Current**, **Select**, or **Create New**, from the **Configuration** drop-down to select the configuration for analysis. These options are discussed next.

Use Current

This option is used to add a new flow simulation project based on the current configuration.

Select

This option is used to create a new flow simulation project based on an existing configuration. You can choose the appropriate configuration from the **Configuration name** drop-down.

Create New

This option is used to create a new configuration based on an existing configuration for a new flow simulation project. When you select this option, you can specify the configuration name in the **Configuration name** edit box.

Choose the **Next** button after specifying the configuration and the name of the project; the **Wizard - Unit System** page is displayed, refer to Figure 4-2. The options in this page are discussed next.

Using the options in this page, you can select the unit system or create a new unit system based on your requirements. These options are discussed next.

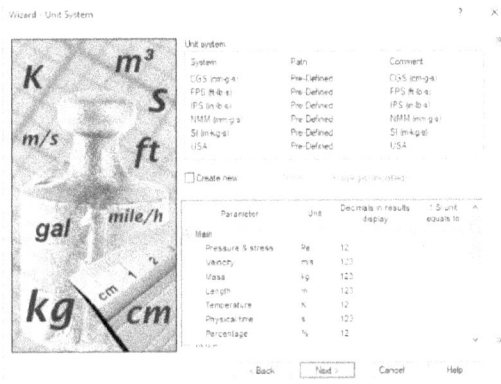

*Figure 4-2 The **Wizard - Unit System** page*

Unit System

In this area, you can specify a unit system from the list of predefined unit systems. The list of predefined unit systems is as follows:

1) CGS (cm-g-s): centimeter gram second
2) FPS (ft-lb-s): foot pound second
3) IPS (in-lb-s): inch pound second
4) NMM (mm-g-s): Newton millimeter kilogram second
5) SI (m-kg-s): Newton meter kilogram second
6) USA

If the default unit system does not match your needs, you can choose a system that more closely matches to your needs and then edit the selected unit system.

To create a modified unit system, select the **Create new** check box and enter the name of the unit system in the **Name** edit box.

You can also change the number of decimal places for displaying analysis results. To do so, click the corresponding cell in the **Decimals in results display** column and select the desired option from the drop-down.

Choose the **Next** button after specifying the unit system; the **Wizard - Analysis Type** page is displayed, refer to Figure 4-3. The options in this page are discussed next.

In this page, you can specify the type of analysis and specific physical feature options for the problem you want to solve with flow simulation.

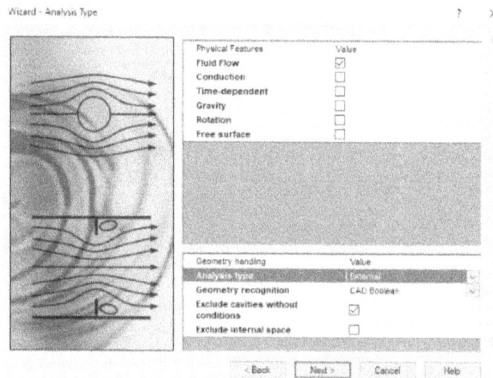

Figure 4-3 The **Wizard - Analysis Type** page

Analysis type

In this area, you can specify whether the flow analysis will be confined to internal or external part of the body.

Internal

Choose this option if the flow analysis is to be confined to solid surfaces on all sides. For example, flow inside the pipeline.

External

Select this option when the flow is simply constrained by the computation domain limits rather than by outer solid surfaces. For example, flow over a solid body.

Consider closed cavities

In this area, you can specify an additional option to reduce the computing resource requirements for complex models with an interior area that is not used in the flow analysis.

Exclude cavities without conditions

Select this check box to exclude close internal spaces that do not have any boundary conditions specified on their surfaces. It can be used to examine both internal and external flows.

Exclude internal space

Select this check box to ignore internal spaces that are closed. This is effective for external flow analysis.

You can also specify the physical feature that is required for your analysis. Select the check box in the **Value** column corresponding to the physical feature to specify the physical feature in the analysis. Some of the physical features that are available in this page are as follows:

a) Conduction
b) Fluid Flow
c) Time-dependent
d) Gravity
e) Rotation
f) Free surface

Choose the **Next** button after specifying the type of analysis and specific physical feature; the **Wizard - Default Fluid** page is displayed, refer to Figure 4-4. The options in this page are discussed next.

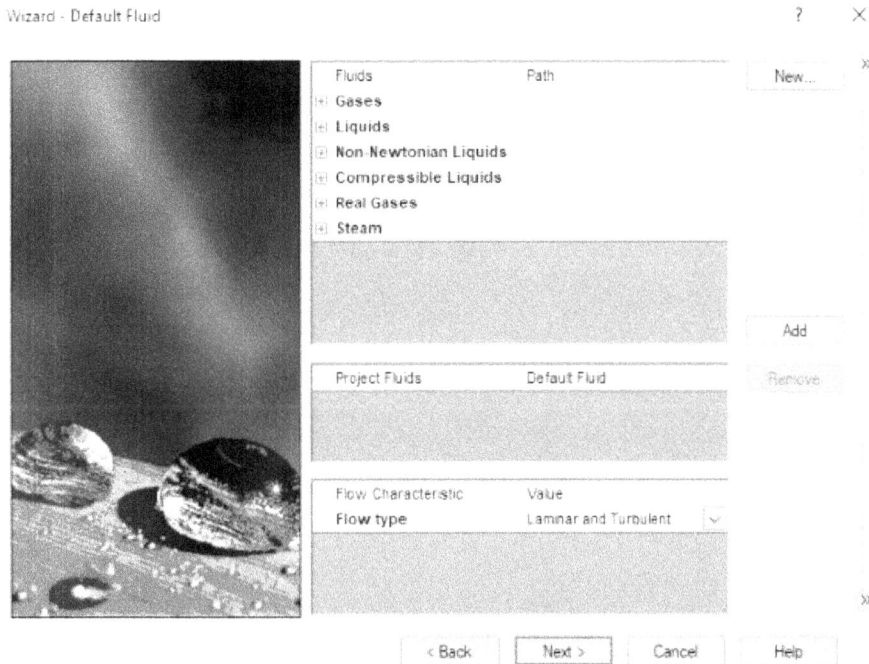

*Figure 4-4 The **Wizard - Default Fluid** page*

Using the options in this page, you can modify the fluid type and also add, remove, or substitute fluid substances. You can select different types of fluids in this dialog box, which are as follows:

a) Gases
b) Liquids
c) Non-Newtonian Liquids
d) Compressible Liquids

e) Real Gases
f) Steam

To add a fluid to the project, click on the ⊞ sign to expand the respective fluid and then select the fluid from the list. Next, choose the **Add** button; the fluid is added under the **Project Fluids** column. To remove a fluid from the **Project Fluids** column, select it and then choose the **Remove** button; the fluid will be removed from the **Project Fluids** column.

To define the type of flow, you can also select laminar or turbulent flow from the **Value** column.

Choose the **Next** button after specifying the fluid type; the **Wizard - Wall Conditions** page is displayed, refer to Figure 4-5. The options in this page are discussed next.

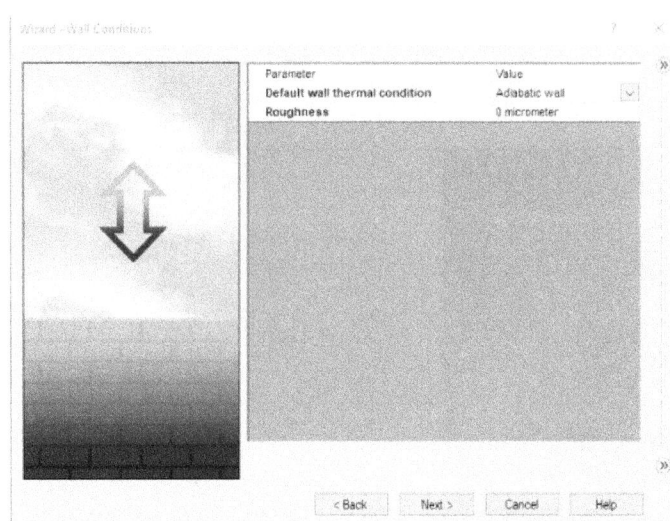

*Figure 4-5   The **Wizard - Wall Conditions** page*

Using the options in this dialog box, you can specify the conditions that are applied to all model walls by default.

In this dialog box, you can select different options from the **Default wall thermal condition** parameter drop-down list and also the roughness value for the wall in the **Roughness** edit box. The options in the **Default wall thermal condition** parameter drop-down list are as follows:

a) Adiabatic wall
b) Heat flux
c) Heat transfer rate
d) Wall temperature

Choose the **Next** button after specifying the wall conditions; the **Wizard - Initial and Ambient Conditions** page is displayed, refer to Figure 4-6. The options in this page are discussed next.

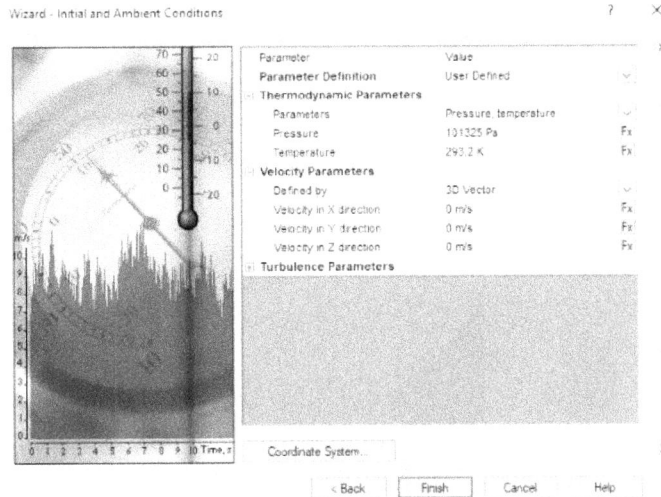

*Figure 4-6 The **Wizard - Initial and Ambient Conditions** page*

Using the options in this page, you can specify the thermodynamic, velocity, turbulence, solid and concentration parameters.

In this page, you can select different options from the **Parameter Definition** drop-down list. The options available in this drop-down list are **User Defined** and **Transferred**. Select the **User Defined** option to define the parameters manually. Select the **Transferred** option to define the parameters from the previous calculation.

Choose the **Finish** button after specifying the parameters for initial conditions; the project is created in the **Flow Simulation Analysis** tree.

TUTORIALS

Tutorial 1

In this tutorial, you will open the model (*c03_tut01*) created in Tutorial 1 of Chapter 3. You can also download this file from *www.cadcim.com* by using the following path:

Textbooks > CAE Simulation > Dassault Systemes > SOLIDWORKS Flow Simulation > Flow Simulation Using SOLIDWORKS 2023 > Input Files > C04_SWFS_inp

You will then create a Flow Simulation project. The model is shown in Figure 4-7.

(Expected time: 20 min)

The following steps are required to complete this tutorial:

a. Open Tutorial 1 of Chapter 3.
b. Save this tutorial in the *c04* folder with a new name.
c. Name the project and add a descriptory comment about it.

d. Set the unit system precision.
e. Select the analysis type.
f. Select the type of fluid.
g. Use default parameters for wall conditions.
h. Use default parameters for initial conditions.

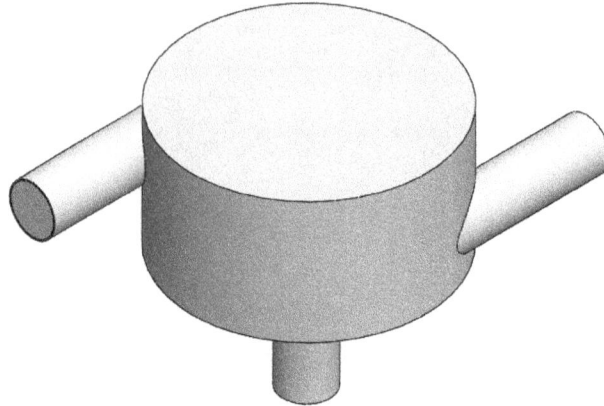

Figure 4-7 Solid model for Tutorial 1

Opening Tutorial 1 of Chapter 3

As the required tutorial is saved in the *c03* folder, you need to select this folder and then open the *c03_tut01.sldprt* document.

1. Start SOLIDWORKS by double-clicking on its shortcut icon on the desktop of your computer.

2. Choose the **Open** button from the Menu Bar to display the **Open** dialog box.

3. Browse to the SOLIDWORKS folder and select the **c03** folder.

4. Select the **c03_tut01.sldprt** document and then choose the **Open** button.

    As the model was saved in the part modeling environment in Chapter 3, it opens in the part modeling environment.

Saving the Document in the c04 Folder

When you open a document from another chapter, it is recommended that you first save the opened document with a new name in the folder of the current chapter to avoid the original document from getting modified.

1. Choose the **Save As** button from the **Save** flyout in the Menu Bar; the **Save As** dialog box is displayed, refer to Figure 4-8.

2. Browse to the **SOLIDWORKS_Flow > Resources** folder and then create a new folder with the name **c04** by using the **Create New Folder** button. Make the **c04** folder as the current folder by double-clicking on it.

3. Enter **c04_tut01** as the new name of the document in the **File** name edit box and then choose the **Save** button to save the document.

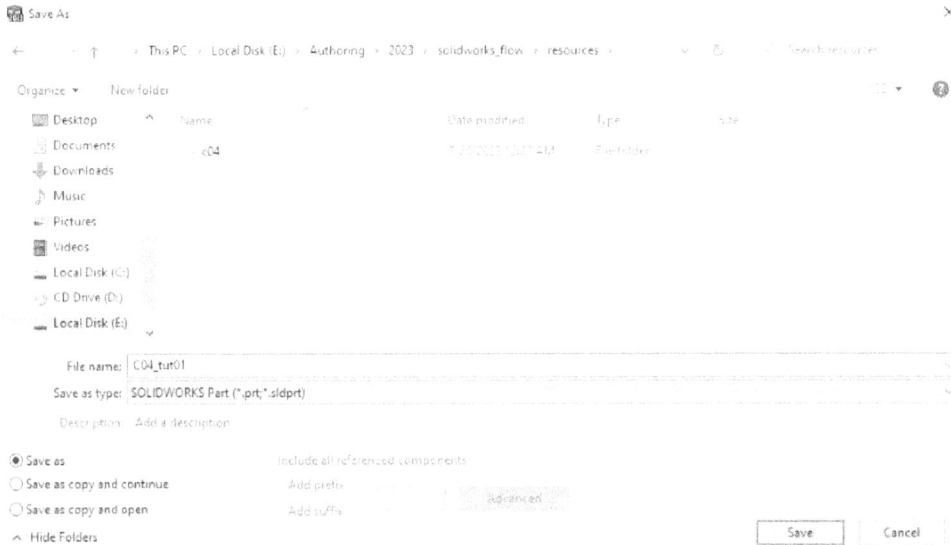

Figure 4-8 The **Save As** *dialog box*

The document is saved with the new name and gets opened in the drawing area.

Create a New Flow Simulation Project

Next, you need to invoke the **Wizard** tool and then create a project in a sequential manner.

1. Choose the **Flow Simulation** tab from the CommandManager to display the **Flow Simulation CommandManager**. Next, choose the **Wizard** button; the **Wizard - Project Name** page is displayed, as shown in Figure 4-9.

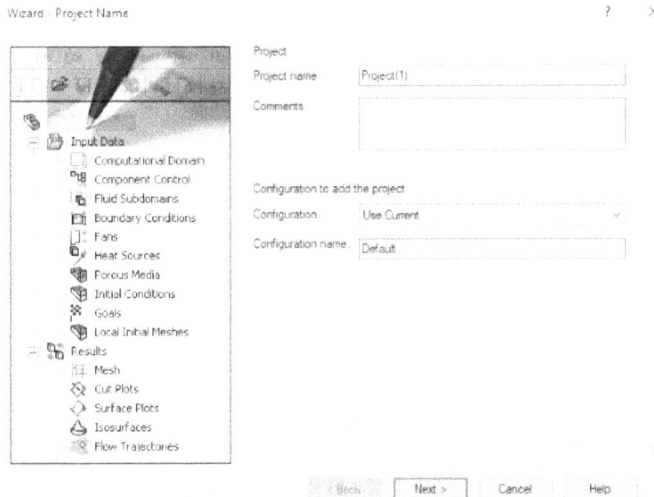

Figure 4-9 The **Wizard - Project Name** *page*

2.   Enter the name **c04_tut_01** in the **Project name** area.

You can also add the comments regarding the project in the **Comments** area.

3.   Enter the name **Internal Flow Simulation** in the **Comments** area.

4.   Ensure that the **Use Current** option is selected in the **Configuration** drop-down. You will notice that **Default** name appears in the **Configuration name** area.

5.   Choose the **Next** button; the **Wizard - Unit System** page is displayed, as shown in Figure 4-10.

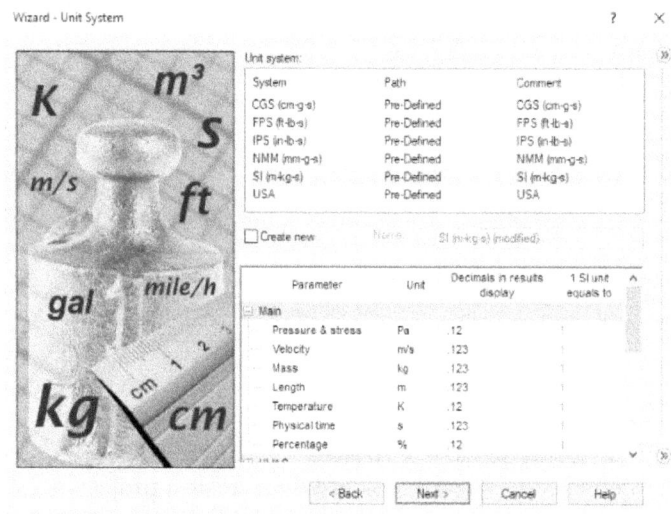

Figure 4-10   The Wizard - Unit System page

6.   Select the **SI (m-kg-s)** system from the **Unit system** area if it is not selected by default. Also ensure that the velocity, mass, length, and temperature have the **.12** option selected in the **Decimals in results display** column.

7.   Choose the **Next** button; the **Wizard - Analysis Type** page is displayed, as shown in Figure 4-11.

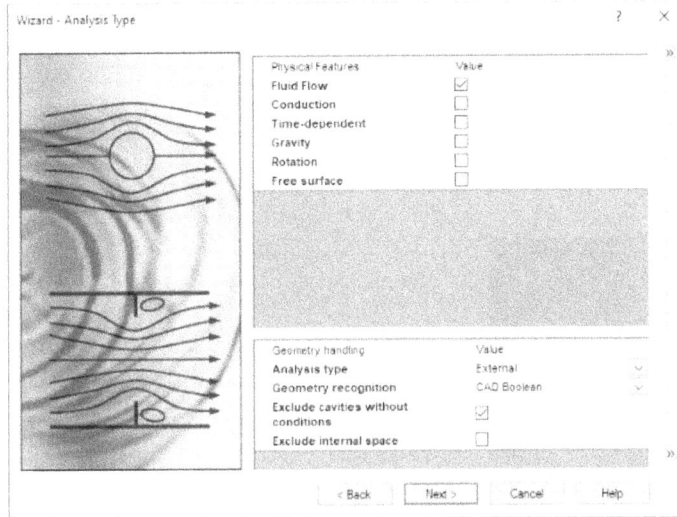

*Figure 4-11   The **Wizard - Analysis Type** page*

8.   Select the **Internal** option from the **Analysis type** drop-down list in the dialog box.

9.   Select the **Exclude cavities without conditions** check box in the **Value** column of the dialog box.

10. Choose the **Next** button; the **Wizard - Default Fluid** page is displayed, as shown in Figure 4-12.

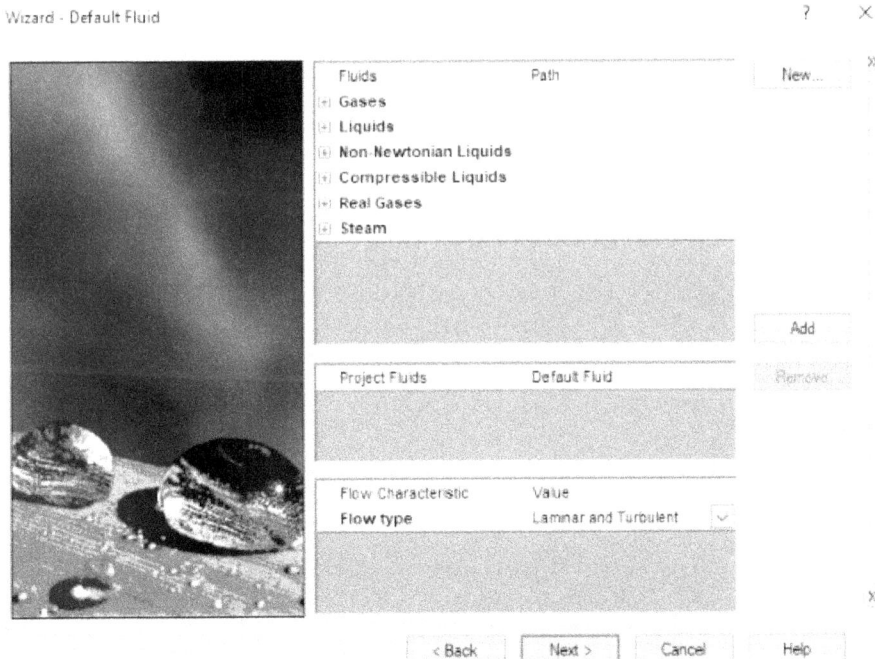

*Figure 4-12   The **Wizard - Default Fluid** page*

11. In the **Fluids** column, click on the + sign on the left of the **Liquids** option to display the list of liquids available, as shown in Figure 4-13.

12. Select **Water** in the **Fluids** column and choose the **Add** button; the **Water (Liquids)** is added in the **Project Fluids** column.

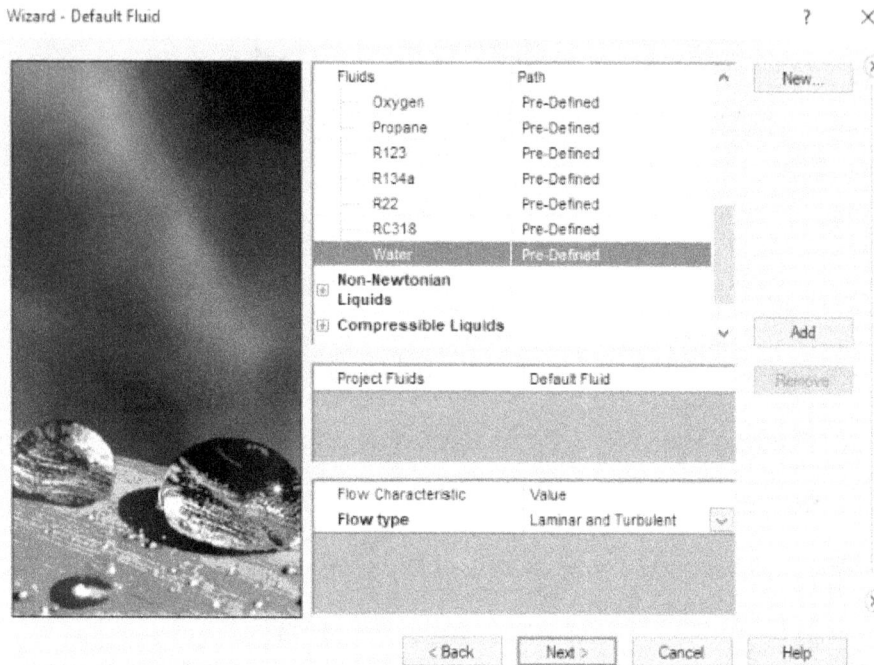

*Figure 4-13 List of liquids in the **Fluids** column*

13. Choose the **Next** button; the **Wizard - Wall Conditions** page is displayed, as shown in Figure 4-14. You can specify the conditions to be applied to the model walls. However, for this tutorial, keep the default parameters as they are.

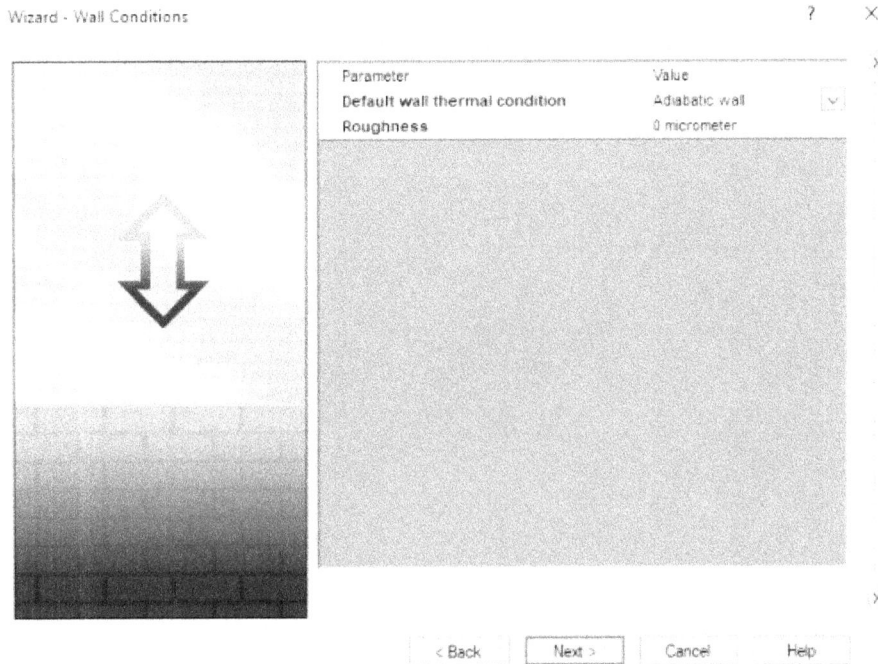

*Figure 4-14   The **Wizard - Wall Conditions** page*

14. Choose the **Next** button; the **Wizard - Initial Conditions** page is displayed, as shown in Figure 4-15. You can specify the initial conditions like **Thermodynamic parameters** or **Velocity Parameters**, but for this tutorial, keep the default values unchanged.

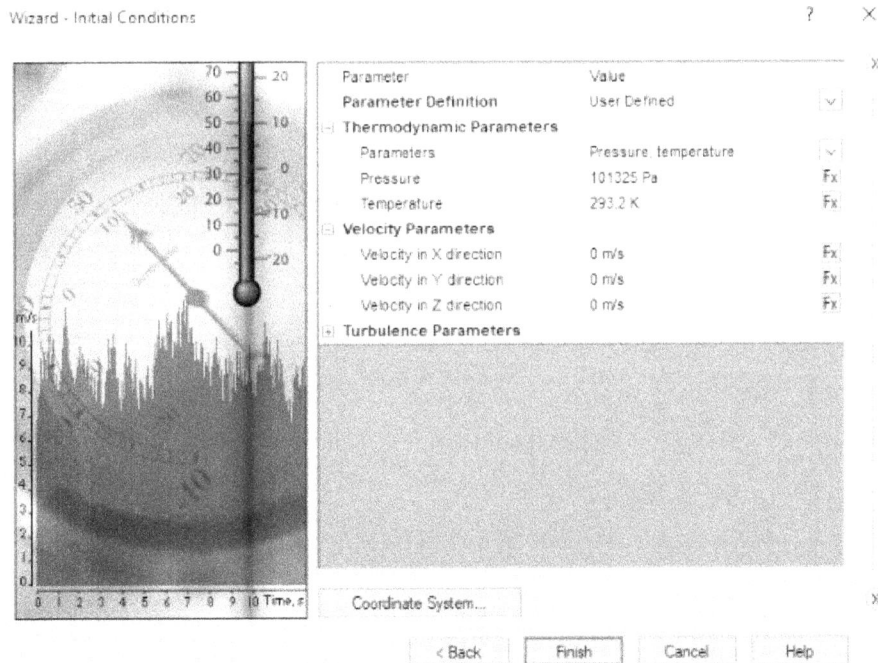

*Figure 4-15   The **Wizard - Initial Conditions** page*

15. Choose the **Finish** button to close the dialog box. You will notice that the **c04_tut_01** project name is added under **Projects in Flow Simulation Analysis** and computational domain is attached to the model, as shown in Figure 4-16.

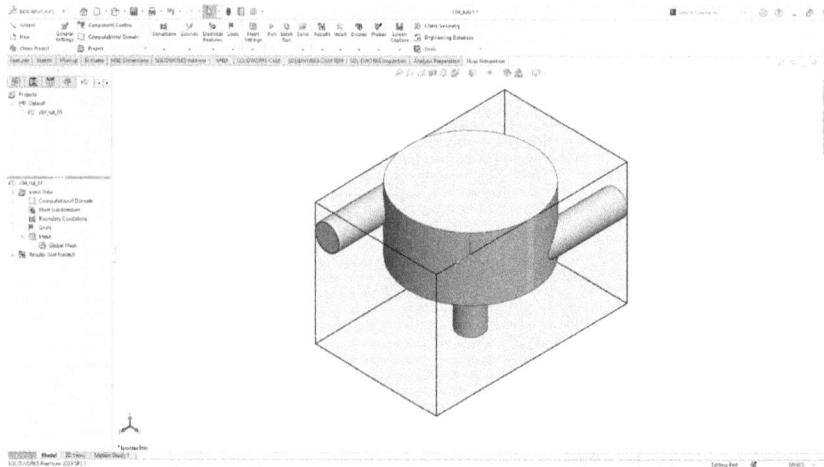

Figure 4-16 *The project added for Flow Simulation analysis with computational domain*

Saving the Model

1. Save the part document with the name *c04_tut01* at the following location: *\SOLIDWORKS_Flow\ resources\c04*.

2. Choose **File > Close** from the SOLIDWORKS menus to close the document.

Tutorial 2

In this tutorial, you will open the model (*c03_tut02*) created in Tutorial 2 of Chapter 3. You can also download this file from *www.cadcim.com* by using the following path:

Textbooks > CAE Simulation > Dassault Systemes > SOLIDWORKS Flow Simulation > Flow Simulation Using SOLIDWORKS 2023 > Input Files > C04_SWFS_inp

You will then create a Flow Simulation project. The model is shown in Figure 4-17.

(Expected time: 30 min)

The following steps are required to complete this tutorial:

a. Open Tutorial 2 of Chapter 3.
b. Save this document in the *c04* folder with a new name.
c. Name the project and add a descriptory comment about it.
d. Set the unit system precision.
e. Select the analysis type.
f. Select the type of fluid.

g.   Use default parameters for wall conditions.

h.   Use default parameters for initial conditions.

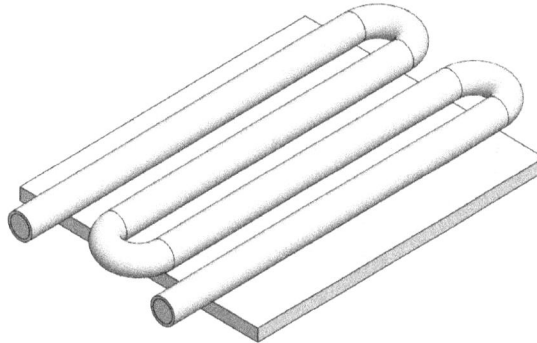

Figure 4-17 *Solid model for Tutorial 2*

Opening Tutorial 2 of Chapter 3

As the required document is saved in the *c03* folder, you need to select this folder and then open the *c03_tut02.sldprt* document.

1.   Start SOLIDWORKS by double-clicking on its shortcut icon on the desktop of your computer.

2.   Choose the **Open** button from the Menu Bar to display the **Open** dialog box.

3.   Browse to the SOLIDWORKS folder and select the **c03** folder.

4.   Select the **c03_tut02.sldprt** document and then choose the **Open** button.

As the model was saved in the part modeling environment in Chapter 3, it opens in the part modeling environment.

Saving the Document in the c04 Folder

When you open a document from another chapter, it is recommended that you save the project with another name in the folder of the current chapter so that the original document is not modified.

1.   Choose the **Save As** button from the **Save** flyout in the Menu Bar; the **Save As** dialog box is displayed, as shown in Figure 4-18.

2.   Browse to the **SOLIDWORKS_Flow > Resources** folder and make the **c04** folder as the current folder by double-clicking on it.

3. Enter **c04_tut02** as the new name of the document in the **File name** edit box and then choose the **Save** button to save the document.

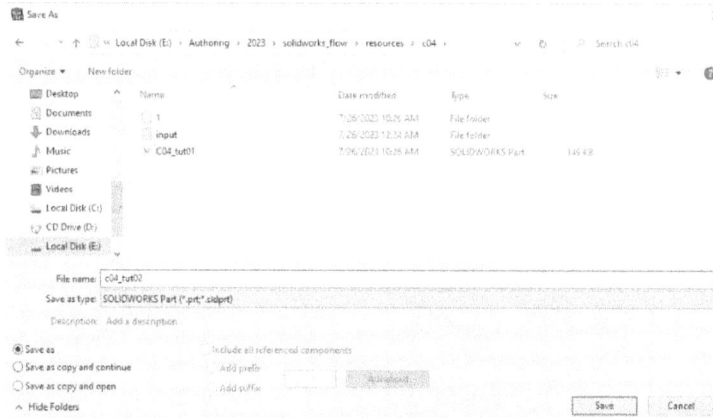

Figure 4-18 The Save As dialog box

4. The document is saved with the new name and gets opened in the drawing area.

Create a New Flow Simulation Project

Next, you need to invoke the **Wizard** tool and create a project in a sequential manner.

1. Choose the **Flow Simulation** tab from the CommandManager to display the **Flow Simulation CommandManager**. Then, choose the **Wizard** button; the **Wizard - Project Name** page is displayed, as shown in Figure 4-19.

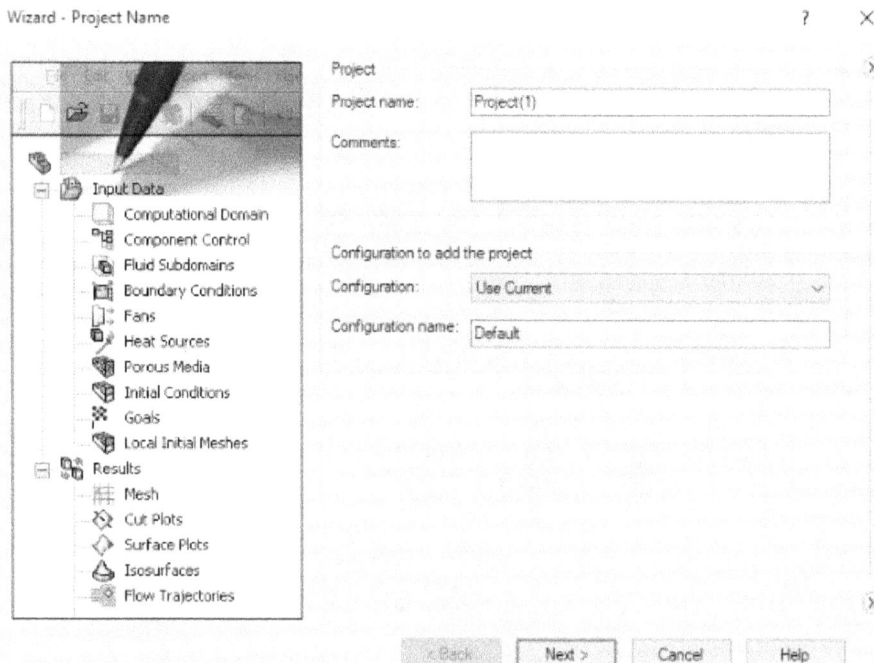

Figure 4-19 The Wizard - Project Name page

2. Enter the name **c04_tut_02** in the **Project name** area.

   You can also add the comments regarding the project in the **Comments** area.

3. Enter the name **Heat Transfer Analysis** in the **Comments** area.

4. Ensure that the **Use Current** option is selected in the **Configuration** drop-down. You will notice that **Default** appears in the **Configuration name** text box.

5. Choose the **Next** button; the **Wizard - Unit System** page is displayed, as shown in Figure 4-20.

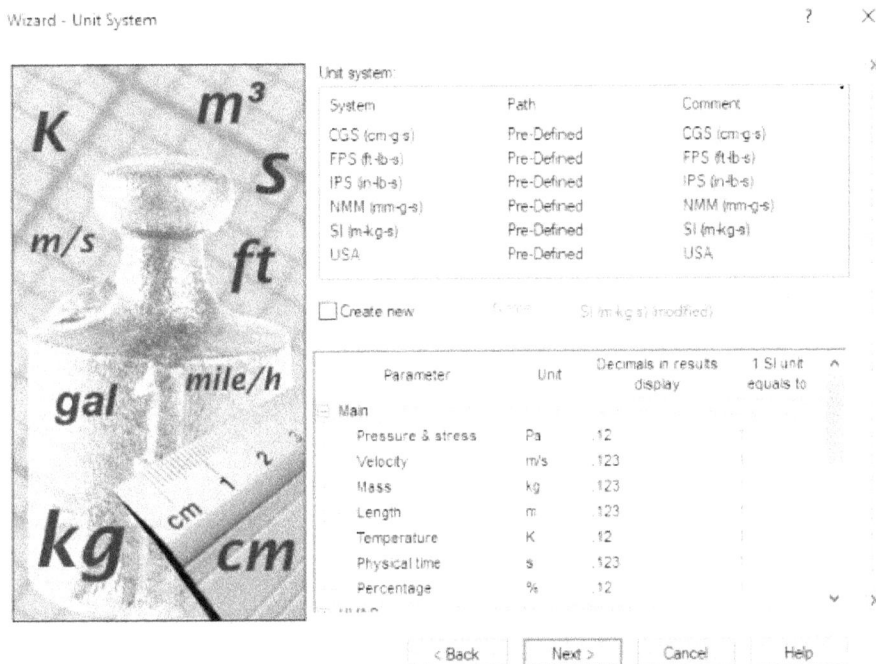

Figure 4-20 The Wizard - Unit System page

6. Select the **SI (m-kg-s)** system from the **Unit System** tab if it is not selected by default. Also ensure that the velocity, mass, length, and temperature have the **.12** option specified in the **Decimals in results display** column.

7. Choose the **Next** button; the **Wizard - Analysis Type** page is displayed, as shown in Figure 4-21.

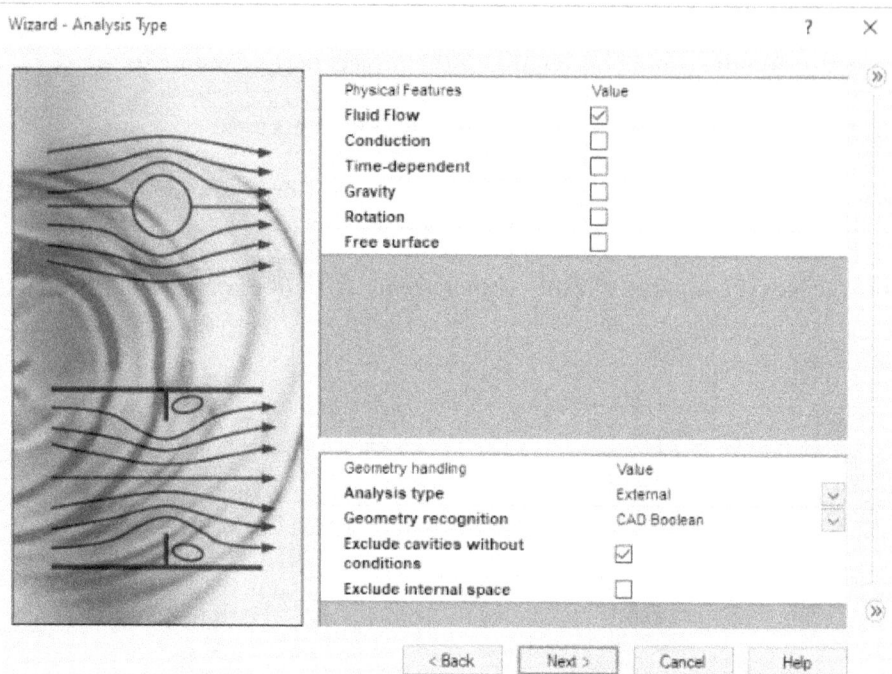

Figure 4-21   The Wizard - Analysis Type page

8. Select the **External** option under the **Analysis type** drop-down in the dialog box.

9. Select the check box in the **Value** column corresponding to the **Conduction** option under the **Physical Features** column.

10. Select the check box in the **Value** column corresponding to the **Gravity** option under the **Physical Features** column.

11. Choose the **Next** button; the **Wizard - Default Fluid** page is displayed, as shown in Figure 4-22.

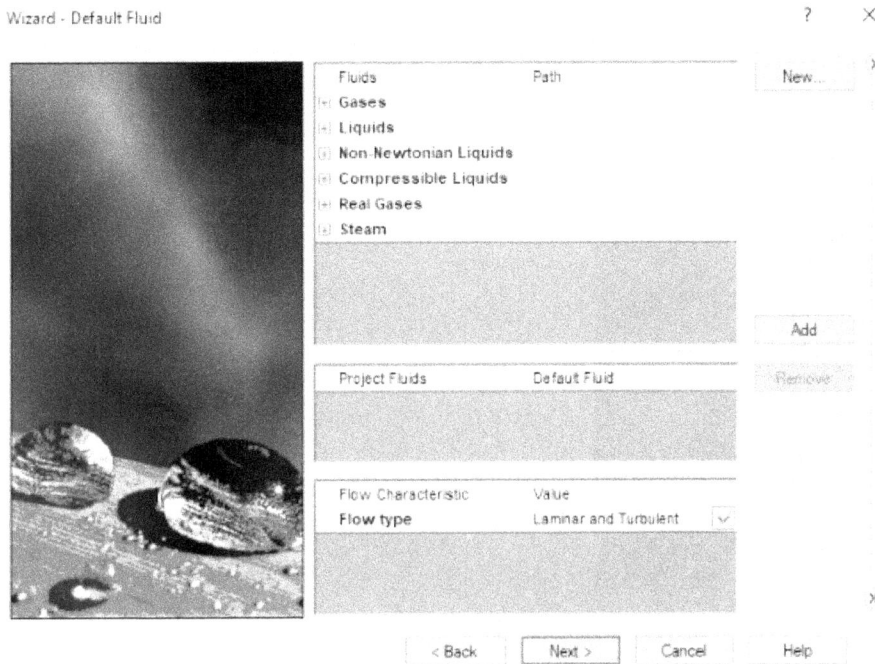

Figure 4-22 *The **Wizard - Default Fluid** page*

12. In the **Fluids** column, expand the **Gases** node to display the list of gases available under it, as shown in Figure 4-23.

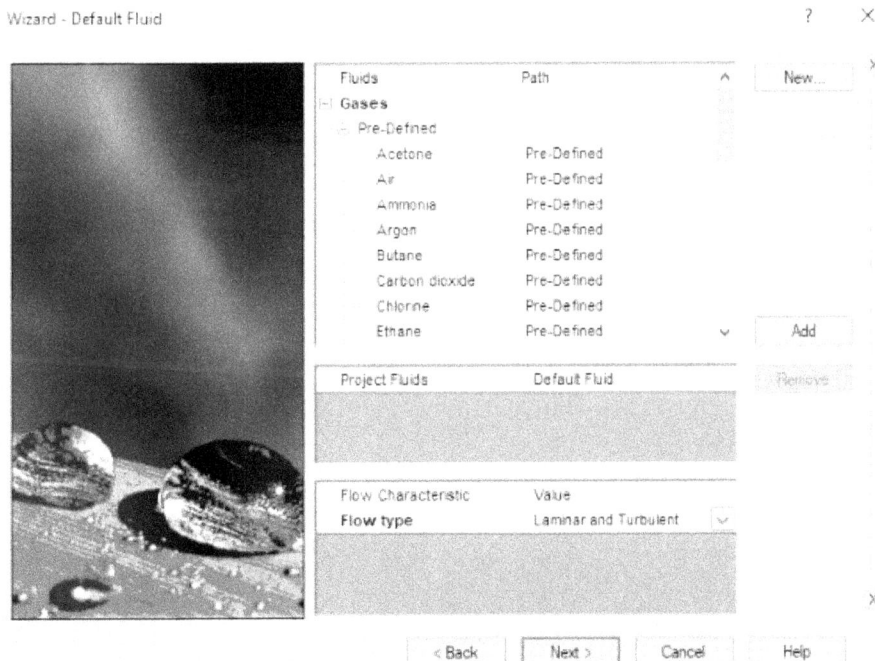

Figure 4-23 *List of gases under the **Gases** node*

13. Select **Air** in the **Fluids** column and choose the **Add** button; the **Air (Gases)** is added in the **Project Fluids** column, as shown in Figure 4-24.

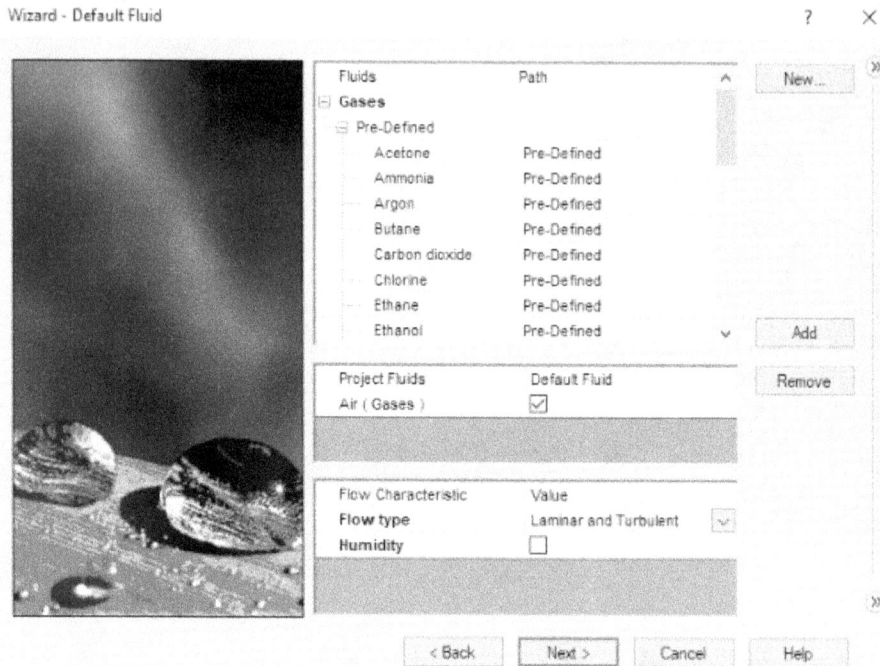

Figure 4-24 *Air added as a gas in the **Project Fluids** column*

14. Click on the ⊞ sign on left of the **Real Gases** option to display the list of real gases available under the **Real Gases** node in the **Fluids** column.

15. Select **Refrigerant R-134a** in the **Fluids** column and then choose the **Add** button; the **Refrigerant R-134a (Real Gases)** is added in the **Project Fluids** column.

16. Choose the **Next** button; the **Wizard - Default Solid** page is displayed, as shown in Figure 4-25.

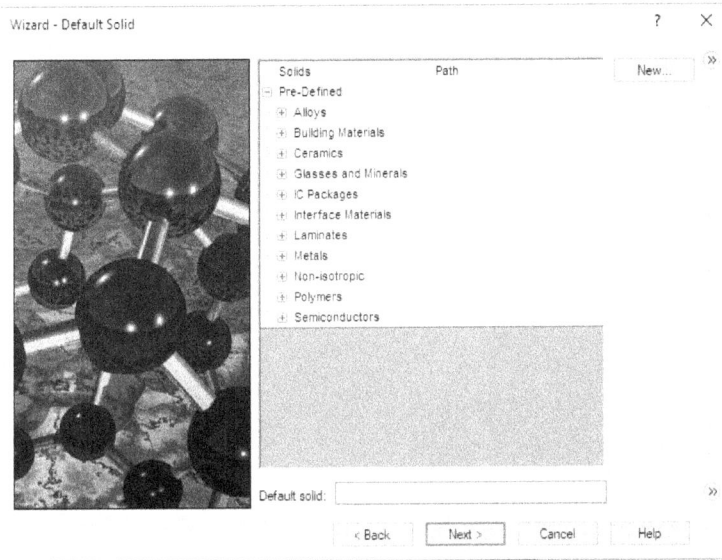

Figure 4-25   *The Wizard - Default Solid page*

17. Click on the + sign on the left of **Alloys** to display the list of alloys available under the **Solids** column.

18. Select **Aluminium 6061** in the **Solids** column; the **Aluminium 6061** is added as the default solid.

19. Choose the **Next** button; the **Wizard - Wall Conditions** page is displayed, as shown in Figure 4-26. You can specify the conditions to be applied to the model walls, but for this tutorial, they will remain the same.

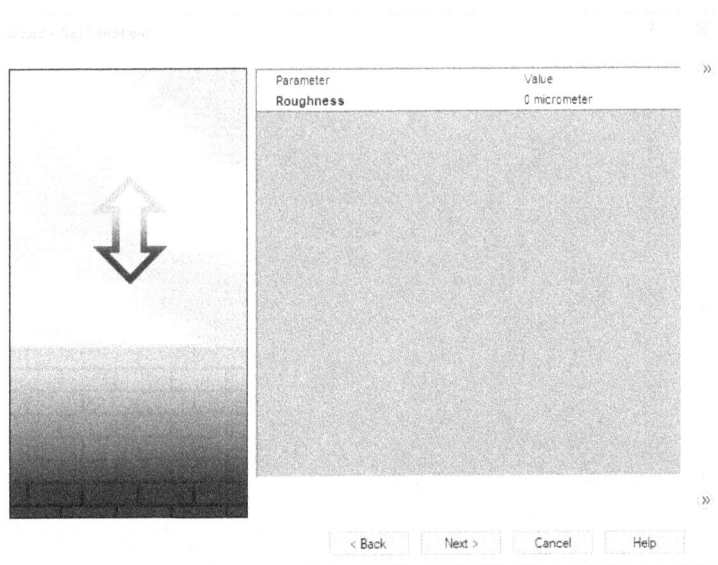

Figure 4-26   *The Wizard - Wall Conditions page*

20. Choose the **Next** button; the **Wizard - Initial and Ambient Conditions** page is displayed, as shown in Figure 4-27.

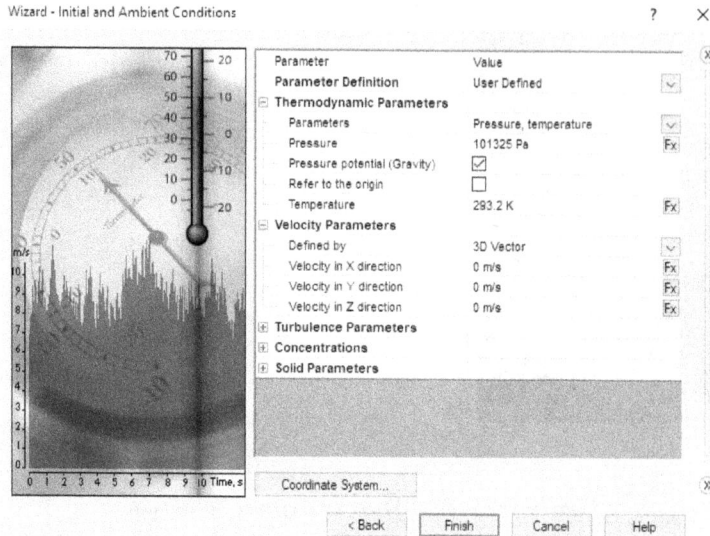

Figure 4-27   The **Wizard - Initial and Ambient Conditions** *page*

21. Click on the ⊕ sign on the left of **Concentrations** to expand it. Enter **0.2** and **0.8** in the **Air** and **Refrigerant R-134a** edit boxes, respectively.

22. Choose the **Finish** button to close the dialog box. You will notice that the **c04_tut_02** project name is added under **Projects** in **Flow Simulation Analysis** and computational domain is attached to the model, as shown in Figure 4-28.

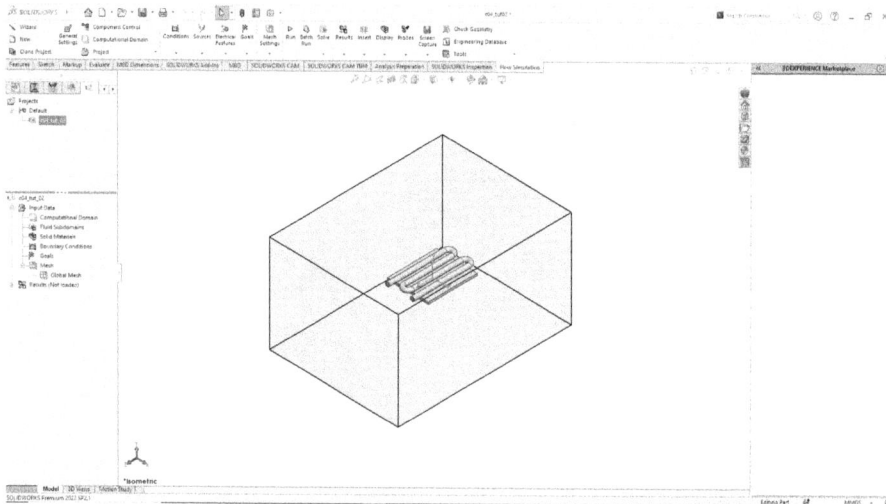

Figure 4-28   The project added for flow simulation analysis with computational domain

Saving the Model

1. Save the part document with the name *c04_tut02* at the following location: *\SOLIDWORKS_ Flow\resources\c04*.

2. Choose **File > Close** from the SOLIDWORKS menus to close the document.

Tutorial 3

In this tutorial, you will open the model (*c03_tut03*) created in Tutorial 3 of Chapter 3. You can also download this file from *www.cadcim.com* by using the following path:

Textbooks > CAE Simulation > Dassault Systemes > SOLIDWORKS Flow Simulation > Flow Simulation Using SOLIDWORKS 2023 > Input Files > C04_SWFS_inp

You will then create a Flow Simulation project. The model is shown in Figure 4-29.

(Expected time: 30 min)

The following steps are required to complete this tutorial:

a. Open Tutorial 3 of Chapter 3.
b. Save this document in the *c04* folder with a new name.
c. Name the project and add a descriptory comment about it.
d. Set the unit system precision.
e. Select the analysis type.
f. Select the type of fluid.
g. Use default parameters for wall conditions.
h. Use default parameters for initial conditions.

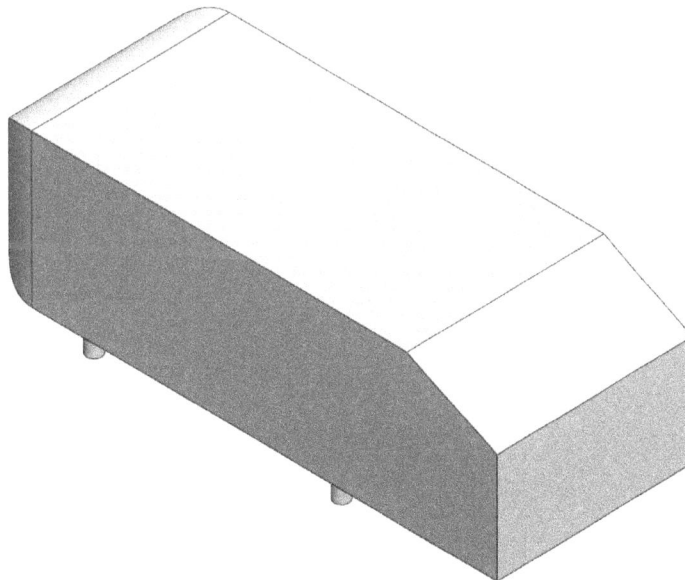

Figure 4-29 *Solid model for Tutorial 3*

Opening Tutorial 3 of Chapter 3

As the required document is saved in the *c03* folder, you need to select this folder and then open the *c03_tut03.sldprt* document.

1. Start SOLIDWORKS by double-clicking on its shortcut icon on the desktop of your computer.

2. Choose the **Open** button from the Menu Bar to display the **Open** dialog box.

3. Browse to the SOLIDWORKS folder and select the **c03** folder.

4. Select the **c03_tut03.sldprt** document and then choose the **Open** button.

    As the model was saved in the part modeling environment in Chapter 3, it opens in the part modeling environment.

Saving the Document in the c04 Folder

When you open a document from another chapter, it is recommended that you save the project with another name in the folder of the current chapter so that the original document is not modified.

1. Choose the **Save As** button from the **Save** flyout in the Menu Bar; the **Save As** dialog box is displayed, as shown in Figure 4-30.

2. Browse to the **SOLIDWORKS_Flow > Resources** folder and make the **c04** folder as the current folder by double-clicking on it.

3. Enter **c04_tut03** as the new name of the document in the **File name** edit box and then choose the **Save** button to save the document.

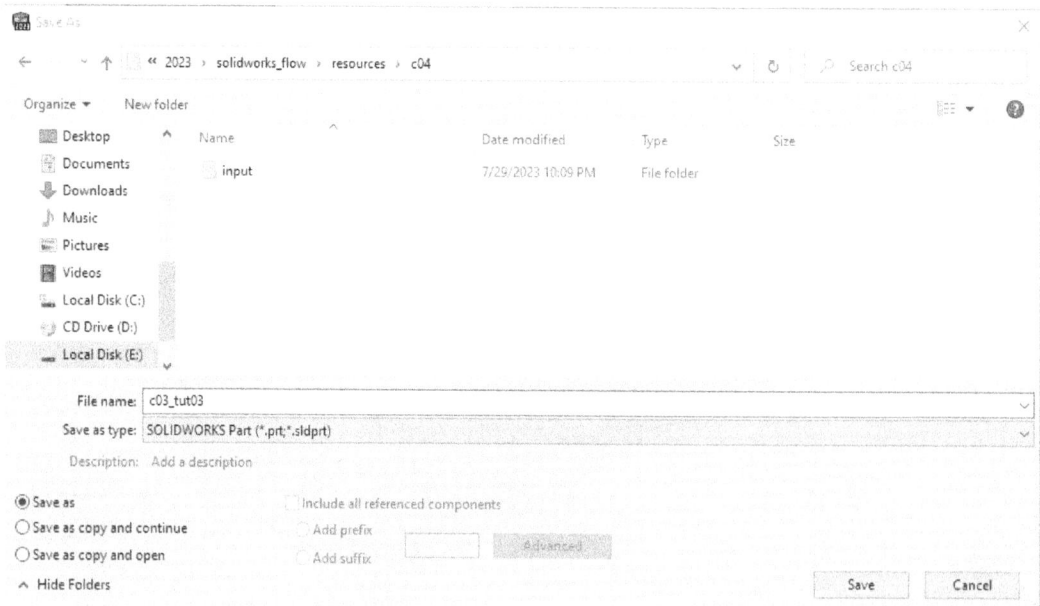

Figure 4-30 *The **Save As** dialog box*

4.   The document is saved with the new name and gets opened in the drawing area.

Create a New Flow Simulation Project

Next, you need to invoke the **Wizard** tool and create a project in a sequential manner.

1.   Choose the **Flow Simulation** tab from the CommandManager to display the **Flow Simulation CommandManager**. Then, choose the **Wizard** button; the **Wizard - Project Name** page is displayed, as shown in Figure 4-31.

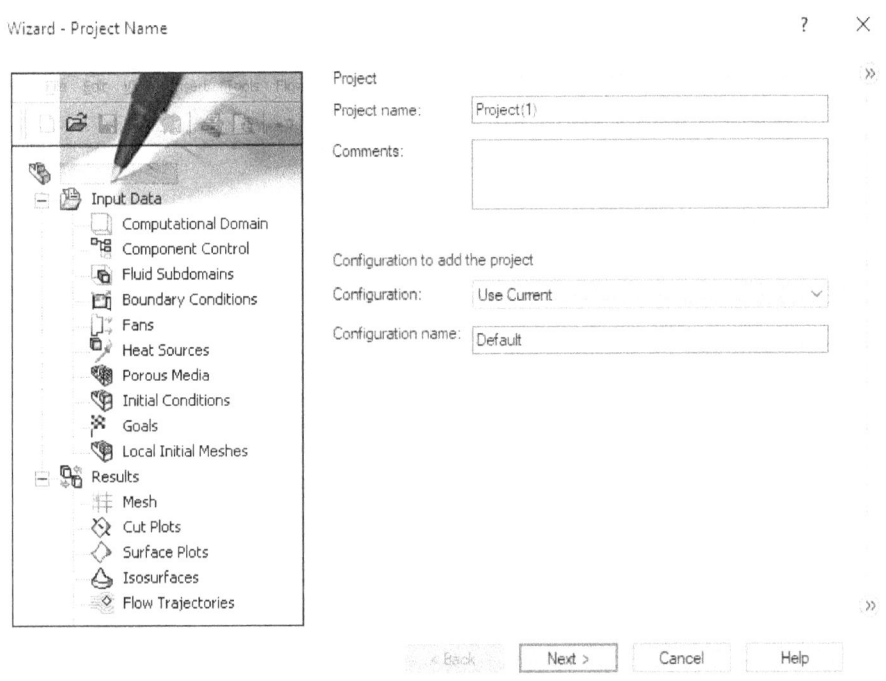

*Figure 4-31   The **Wizard - Project Name** page*

2.   Enter the name **c04_tut_03** in the **Project name** area.

You can also add the comments regarding the project in the **Comments** area.

3.   Enter the name **External Flow Simulation** in the **Comments** area.

4.   Ensure that the **Use Current** option is selected in the **Configuration** drop-down. You will notice that **Default** appears in the **Configuration name** text box.

5.   Choose the **Next** button; the **Wizard - Unit System** page is displayed, as shown in Figure 4-32.

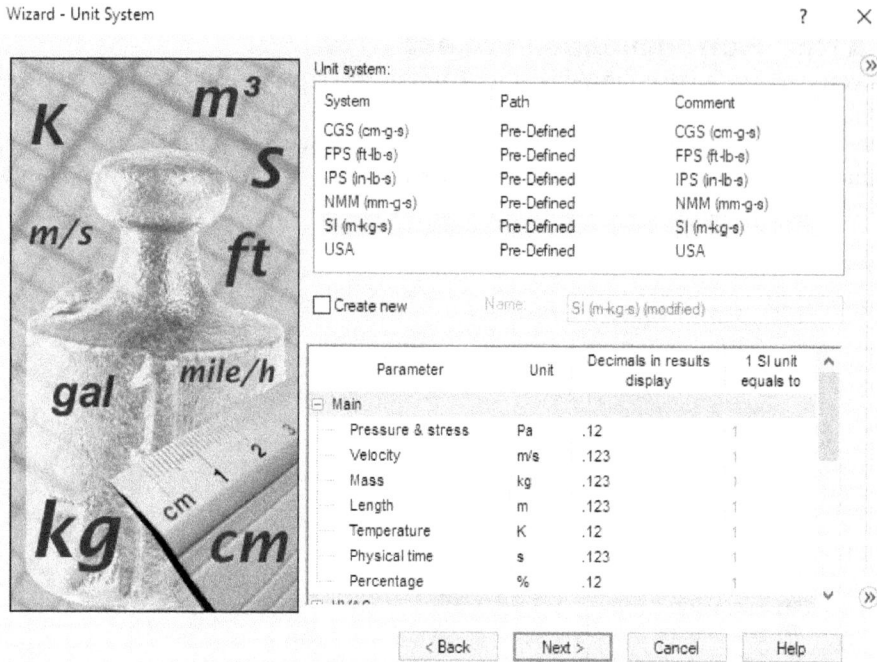

*Figure 4-32   The **Wizard - Unit System** page*

6. Select the **SI (m-kg-s)** system from the **Unit System** tab if it is not selected by default. Also ensure that **.12** is specified as the velocity, mass, length, and temperature in the **Decimals in results display** column.

7. Choose the **Next** button; the **Wizard - Analysis Type** page is displayed, as shown in Figure 4-33.

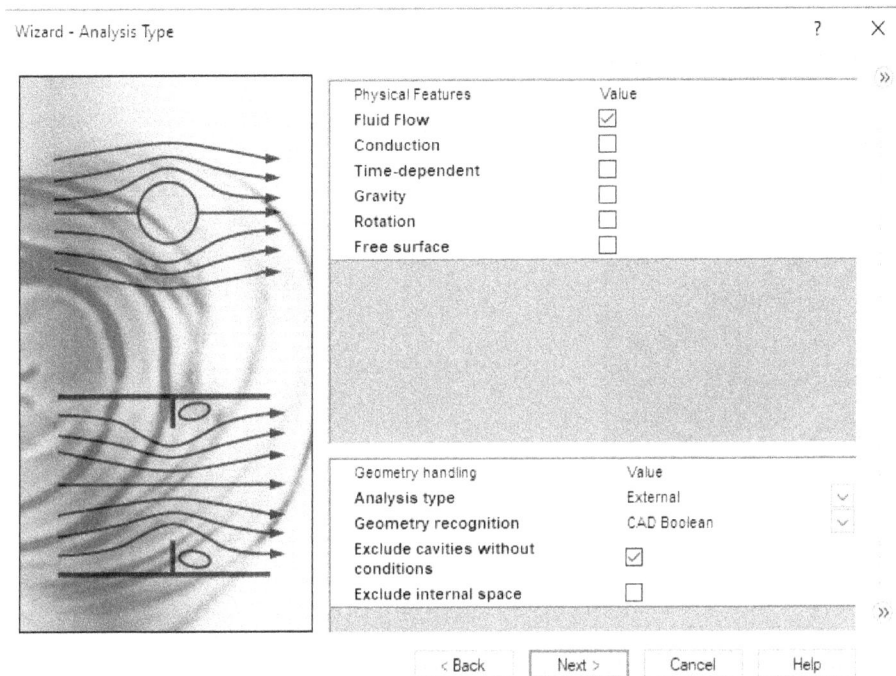

Figure 4-33 The Wizard - Analysis Type page

8. Select the **External** option from the **Analysis type** drop-down in the dialog box.

9. In the **Value** column, select the check box corresponding to the **Fluid Flow** option, if this check box is not selected.

10. Choose the **Next** button; the **Wizard - Default Fluid** page is displayed, as shown in Figure 4-34.

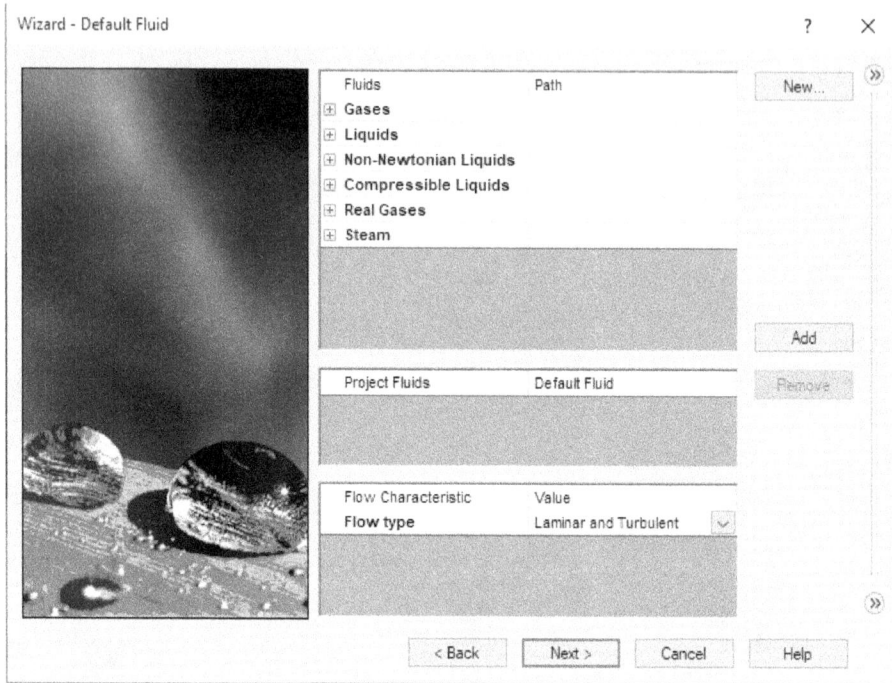

*Figure 4-34 The **Wizard - Default Fluid** page*

11. Click on ⊞ sign placed on the left of the **Gases** option to display the list of gases available under the **Gases** node in the **Fluids** column, as shown in Figure 4-35.

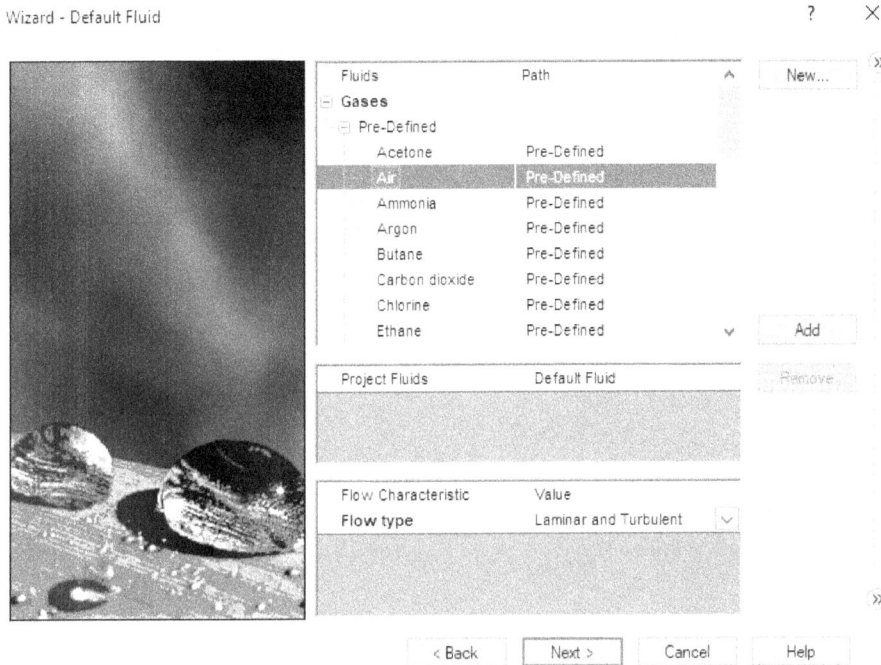

Figure 4-35 *List of gases in the* ***Fluids*** *column*

12. Select **Air** in the **Fluids** column and choose the **Add** button; the **Air (Gases)** is added in the **Project Fluids** column, as shown in Figure 4-36.

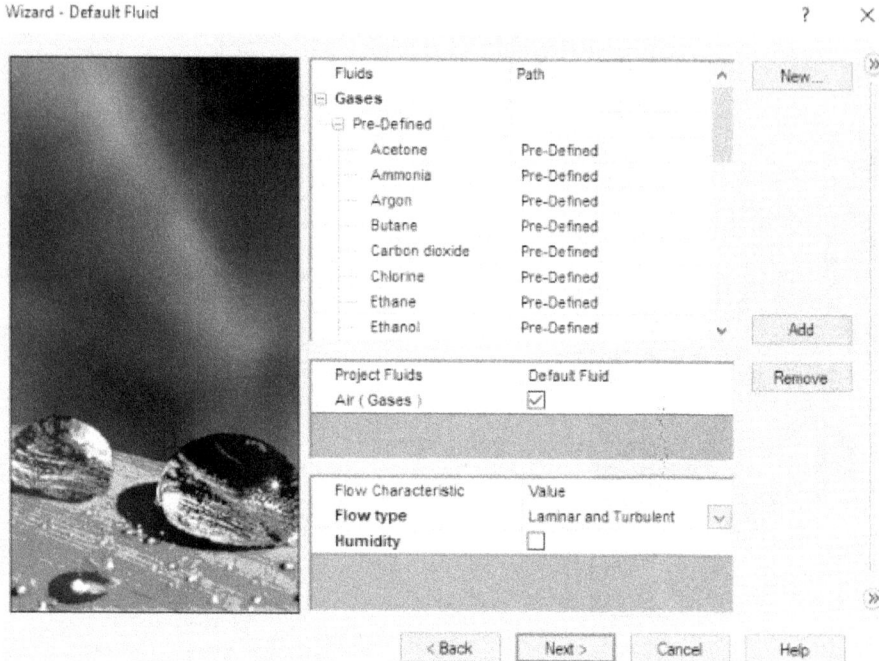

Figure 4-36 *Gas added in the **Project Fluids** column*

13. Choose the **Next** button; the **Wizard - Wall Conditions** page is displayed, as shown in Figure 4-37. You can specify the conditions to be applied to the model walls, but for this tutorial, they will remain the same.

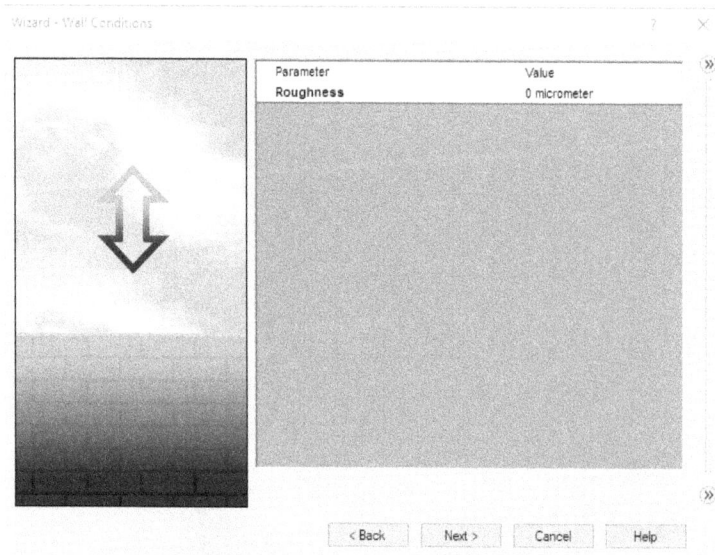

Figure 4-37 *The **Wizard - Wall Conditions** page*

14. Choose the **Next** button; the **Wizard - Initial and Ambient Conditions** page is displayed, as shown in Figure 4-38.

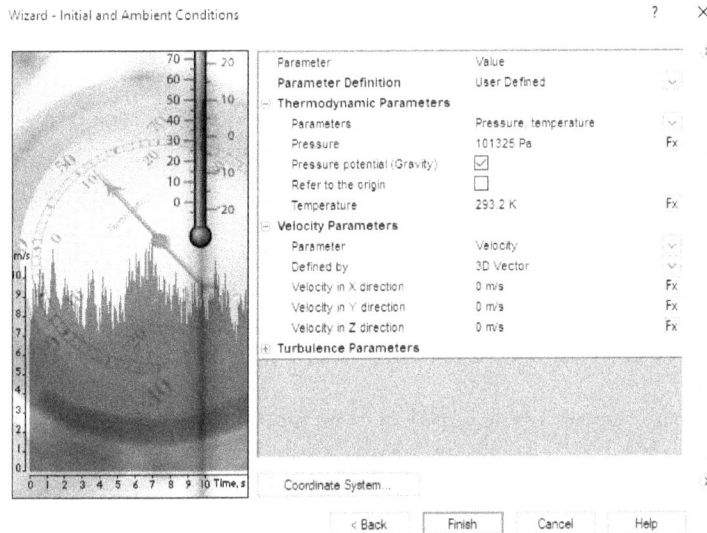

Figure 4-38 The Wizard - Initial and Ambient Conditions page

15. Enter **40** in the **Value** column corresponding to the **Velocity in X direction** parameter in the **Parameter** column.

16. Choose the **Finish** button to close the dialog box. You will notice that the **c04_tut_03** project name is added under **Projects** in **Flow Simulation Analysis** and computational domain is attached to the model, as shown in Figure 4-39.

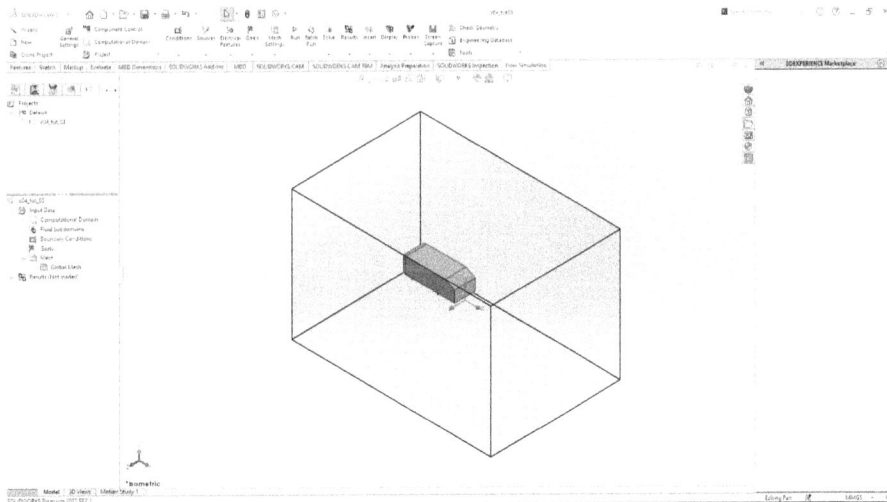

Figure 4-39 The project added for analysis with computational domain

Saving the Model

1. Save the part document with the name *c04_tut03* at the following location: *SOLIDWORKS_ Flow\resources\c04*.

2. Choose **File > Close** from the SOLIDWORKS menus to close the document.

Tutorial 4

In this tutorial, you will open the model (*c03_tut04*) created in Tutorial 4 of Chapter 3. You can also download this file from *www.cadcim.com* by using the following path:

Textbooks > CAE Simulation > Dassault Systemes > *SOLIDWORKS Flow Simulation* > Flow Simulation *U*sing SOLIDWORKS 2023 > *Input Files* > *C04_SWFS_inp*

You will then create a Flow Simulation project. The model is shown in Figure 4-40.

(Expected time: 30 min)

The following steps are required to complete this tutorial:

a. Open Tutorial 4 of Chapter 3.
b. Save this document in the *c04* folder with a new name.
c. Name the project and add a descriptory comment about it.
d. Set the unit system precision.
e. Select the analysis type.
f. Select the type of fluid.
g. Use default parameters for wall conditions.
h. Use default parameters for initial conditions.

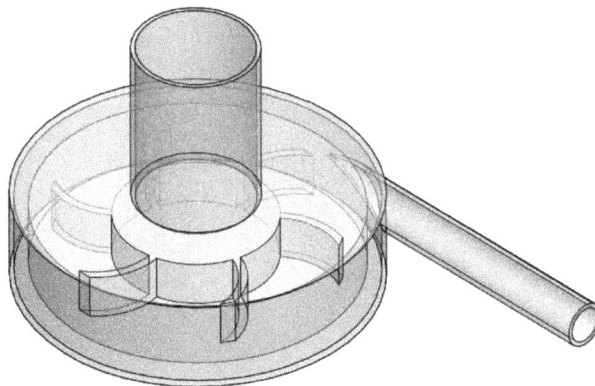

Figure 4-40 *Solid model for Tutorial 4*

Opening Tutorial 4 of Chapter 3

As the required document is saved in the *c03* folder, you need to select this folder and then open the *c03_tut04.sldprt* document.

1. Start SOLIDWORKS by double-clicking on its shortcut icon on the desktop of your computer.

2. Choose the **Open** button from the Menu Bar to display the **Open** dialog box.

3. Browse to the SOLIDWORKS folder and select the **c03** folder.

4. Select the **c03_tut04.sldprt** document and then choose the **Open** button.

    As the model was saved in the part modeling environment in Chapter 3, it opens in the part modeling environment.

Saving the Document in the c04 Folder

When you open a document from another chapter, it is recommended that you save the project with another name in the folder of the current chapter so that the original document is not modified.

1. Choose the **Save As** button from the **Save** flyout in the Menu Bar; the **Save As** dialog box is displayed, as shown in Figure 4-41.

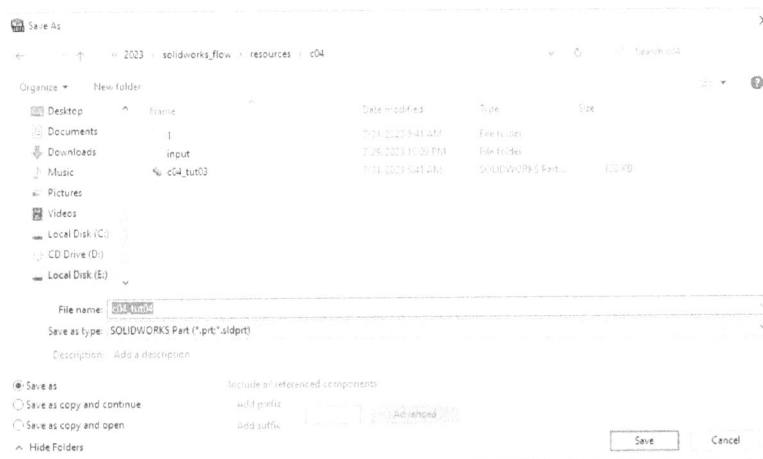

Figure 4-41 The Save As dialog box

2. Browse to the **SOLIDWORKS_Flow > Resources** folder and make the **c04** folder as the current folder by double-clicking on it.

3. Enter **c04_tut04** as the new name of the document in the **File name** edit box and then choose the **Save** button to save the document.

4. The document is saved with the new name and gets opened in the drawing area.

Create a New Flow Simulation Project

Next, you need to invoke the **Wizard** tool and then create a project in a sequential manner.

1. Choose the **Flow Simulation** tab from the CommandManager to display the **Flow Simulation CommandManager**. Then, choose the **Wizard** button; the **Wizard - Project Name** page is displayed, as shown in Figure 4-42.

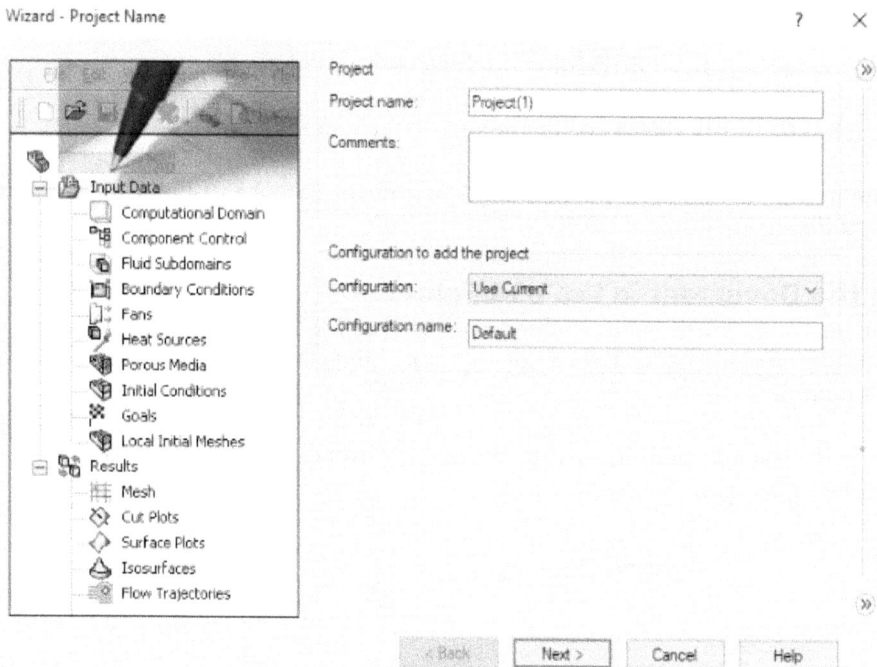

*Figure 4-42 The **Wizard - Project Name** page*

2. Enter the name **c04_tut_04** in the **Project name** area.

    You can also add the comments regarding the project in the **Comments** area.

3. Enter the name **CP Analysis** in the **Comments** area.

4. Ensure that the **Use Current** option is selected in the **Configuration** drop-down. You will notice that **Default** appears in the **Configuration name** text box.

5. Choose the **Next** button; the **Wizard - Unit System** page is displayed, as shown in Figure 4-43.

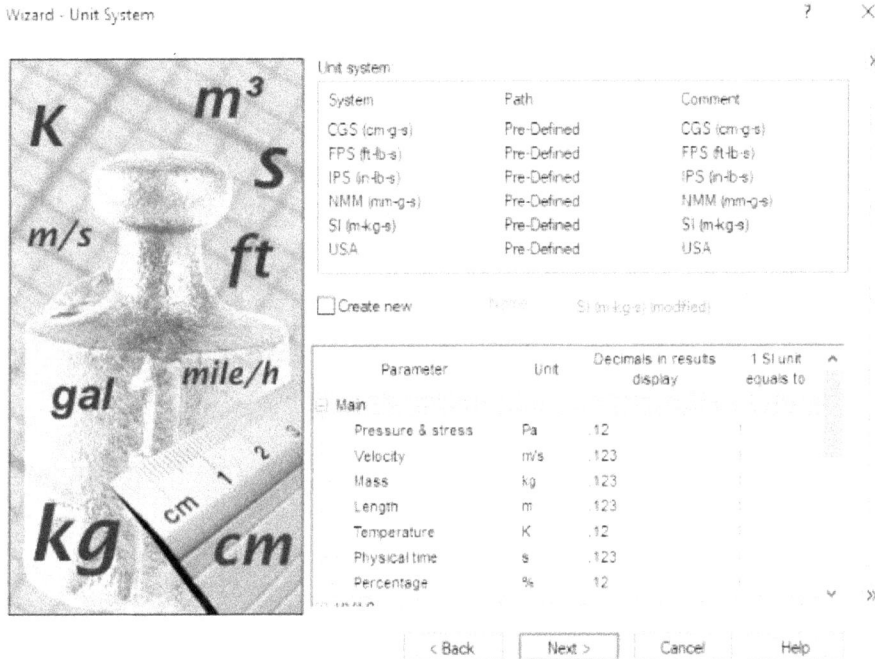

Figure 4-43 The *Wizard - Unit System* page

6. Select the **SI (m-kg-s)** system from the **Unit Sys**tem tab if it is not selected by default. Also ensure that the velocity, mass, length, and temperature have the **.12** option selected in the **Decimals in results display** column.

7. Choose the **Next** button; the **Wizard - Analysis Type** page is displayed, as shown in Figure 4-44.

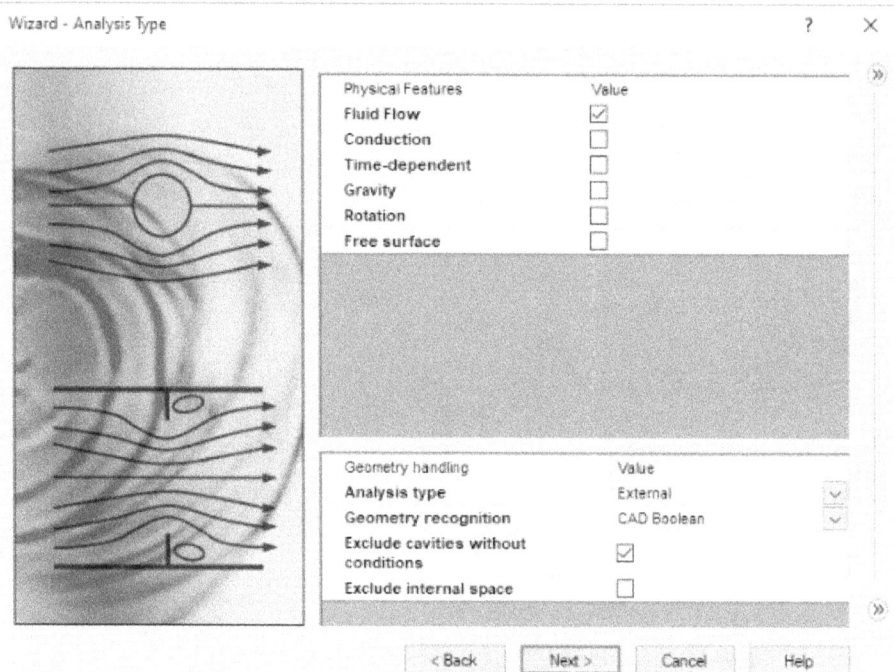

Figure 4-44 The **Wizard - Analysis Type** *page*

8. Select the **Internal** option in the **Value** column for the **Analysis type** feature in the dialog box.

9. Select the **Value** column check box corresponding to the **Rotation** option in the **Physical Features** column; the **Type** drop-down becomes available. Ensure that **Local region(s) (Averaging)** option is selected under the **Type** drop-down.

10. Choose the **Next** button; the **Wizard - Default Fluid** page is displayed, as shown in Figure 4-45.

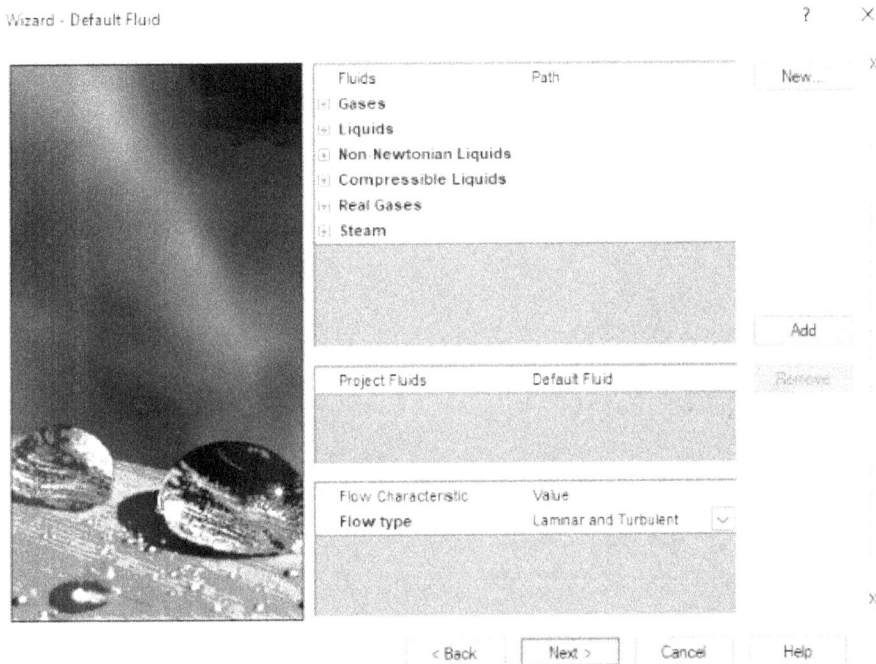

*Figure 4-45   The **Wizard - Default Fluid** page*

11. Click on the ⊞ sign placed adjacent to the **Liquids** node to display the list of gases available under it in the **Fluids** column, as shown in Figure 4-46.

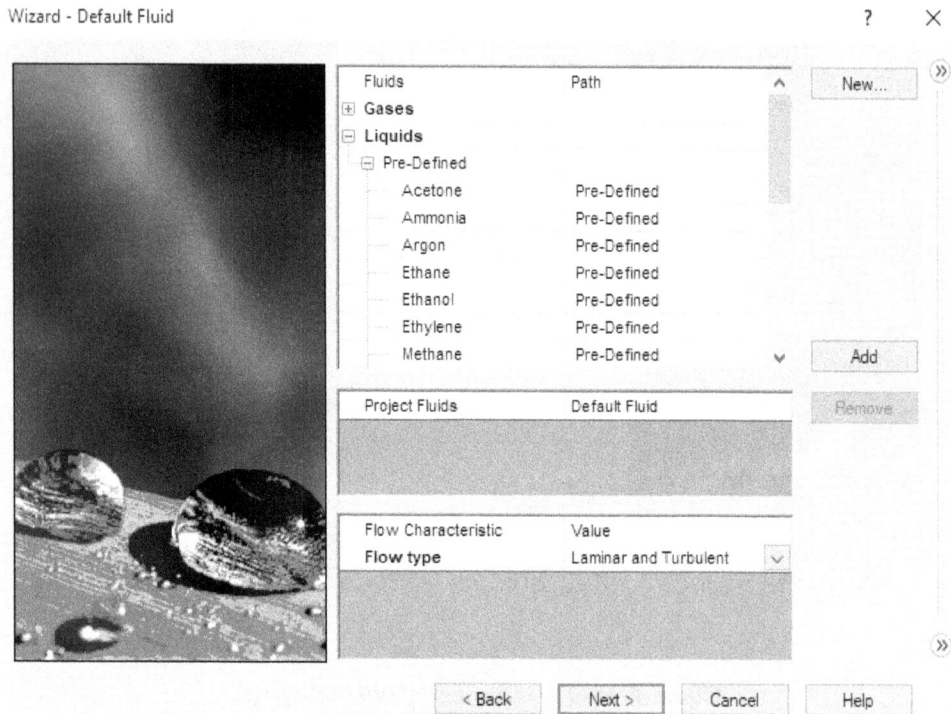

*Figure 4-46 List of liquids in the **Fluids** column*

12. Select **Water** in the **Fluids** column and choose the **Add** button; the **Water (Liquids)** is added in the **Project Fluids** column, as shown in Figure 4-47.

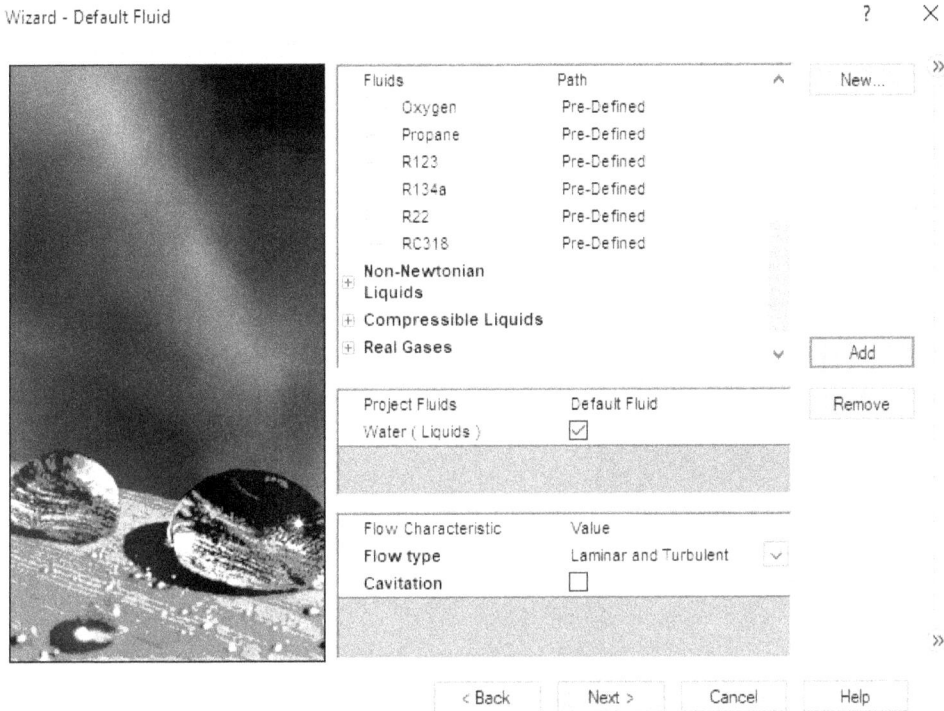

*Figure 4-47 Liquid added in the **Project Fluids** column*

13. Choose the **Next** button; the **Wizard - Wall Conditions** page is displayed, as shown in Figure 4-48. You can specify the conditions to be applied to the model walls, but for this tutorial, they will remain the same.

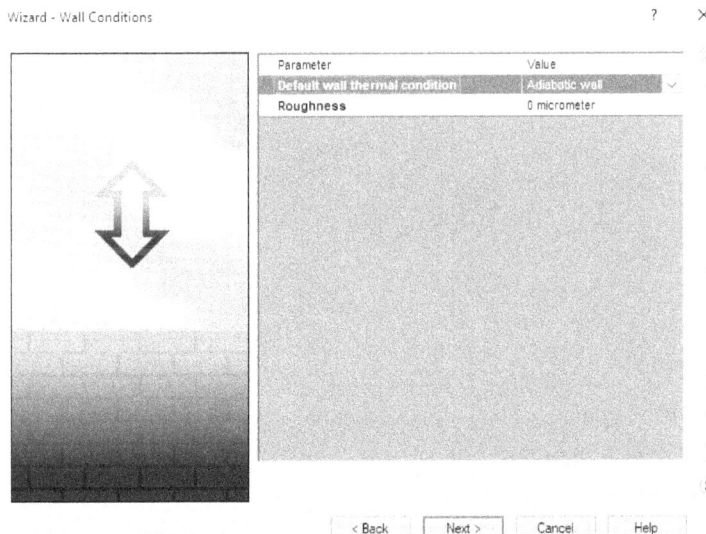

*Figure 4-48 The **Wizard - Wall Conditions** page*

14. Choose the **Next** button; the **Wizard - Initial and Ambient Conditions** page is displayed, as shown in Figure 4-49.

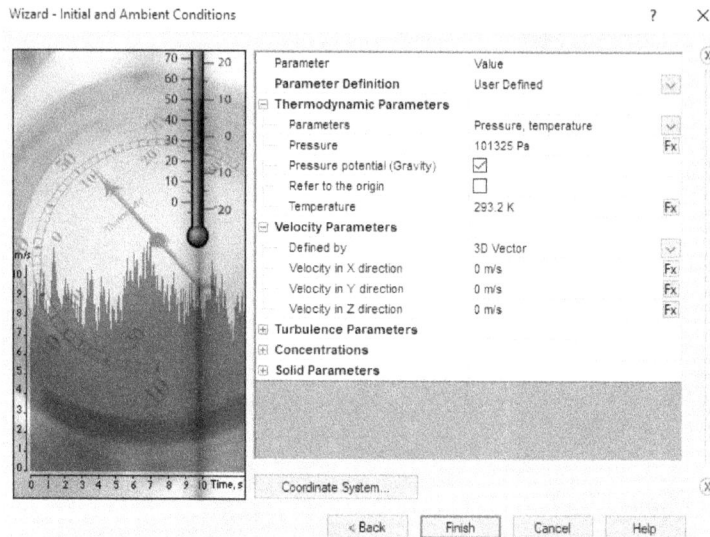

*Figure 4-49   The **Wizard - Initial and Ambient Conditions** page*

15. Choose the **Finish** button to close the dialog box. You will notice that the **c04_tut_04** project name is added under **Projects** in **Flow Simulation Analysis** and the computational domain is attached to the model, as shown in Figure 4-50.

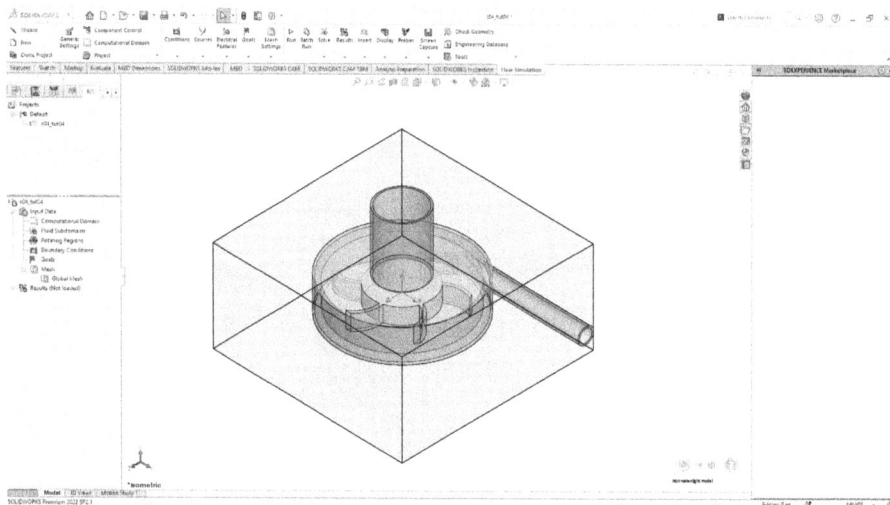

Figure 4-50   The project added for flow simulation analysis with computational domain

Saving the Model

1.   Save the part document with the name *c04_tut04* at the following location: *\SOLIDWORKS_Flow\resources\c04*.

2. Choose **File > Close** from the SOLIDWORKS menus to close the document.

Self-Evaluation Test

Answer the following questions and then compare them to those given at the end of this chapter:

1. The _____ tool is used to create a project in SOLIDWORKS Flow Simulation.

2. The _____ option is used to add a new flow simulation project according to the current configuration.

3. The _____ area is used to select a unit system from the predefined unit systems.

4. Select the _____ radio button to confine the flow to solid surfaces on all sides.

5. You can select the steam as a fluid for a project. (T/F)

6. You can customize the unit system for a project. (T/F)

Review Questions

Answer the following questions:

1. Which of the following pages is displayed when you choose the **Wizard** tool from the **Flow Simulation CommandManager**?

   (a) **Wizard - Initial Conditions**   (b) **Wizard - Default Fluid**
   (c) **Wizard - Project Name**   (d) None of these

2. In the **Wizard - Default Fluid** page, which of the following fluids is available?

   (a) **Gases**   (b) **Liquids**
   (c) **Steam**   (d) All of these

3. In the **Wizard - Analysis Type** page, which of the following physical features is available?

   (a) **Gravity**   (b) **Free Surface**
   (c) **Rotation**   (d) All of these

4. Select the **Internal** radio button when the flow is simply constrained to the computation domain limits rather than to outer solid surfaces. (T/F)

5. You can specify the roughness value for the wall in a project. (T/F)

EXERCISES

Exercise 1

In this exercise, you will open the model (*c04_exr01*) created in Exercise 1 of Chapter 4. You can also download this file from *www.cadcim.com* by using the following path:

Textbooks > CAE Simulation > Dassault Systemes > SOLIDWORKS Flow Simulation > Flow Simulation Using SOLIDWORKS 2023 > Input Files > C04_SWFS_inp

You will then create a Flow Simulation project. The model is shown in Figure 4-51. The project has the following settings:

Project name: Flow around a sphere
Unit System: SI (m-kg-s)
Analysis type: External
Project Fluid: Air
Velocity in X-direction: 0.05 m/s

(Expected time: 10 min)

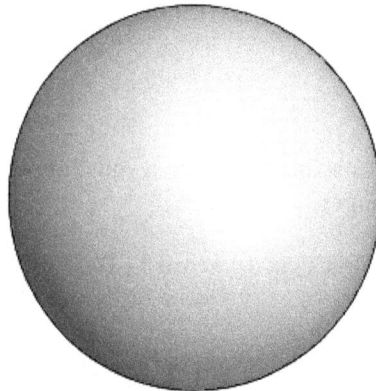

Figure 4-51 Solid model for Exercise 1

Exercise 2

In this exercise, you will open the model (*c04_exr02*) created in Exercise 2 of Chapter 4. You can also download this file from *www.cadcim.com* by using the following path:

Textbooks > CAE Simulation > Dassault Systemes > SOLIDWORKS Flow Simulation > Flow Simulation Using SOLIDWORKS 2023 > Input Files > C04_SWFS_inp

You will then create a Flow Simulation project. The model is shown in Figure 4-52. The project has the following settings:

Project name: Flow around a cylinder
Unit System: SI (m-kg-s)
Analysis type: External
Project Fluid: Air
Velocity in X-direction: 0.08 m/s

(Expected time: 10 min)

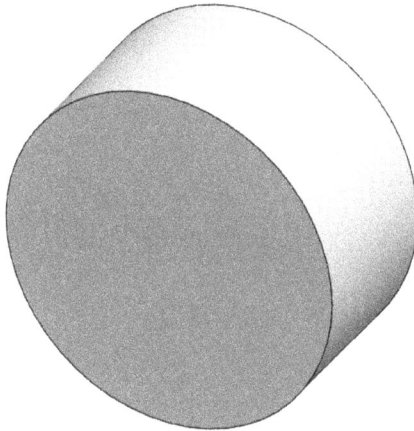

Figure 4-52 *Solid model for Exercise 2*

Answers to Self-Evaluation Test
1. Wizard, 2. Use Current, 3. Unit System, 4. Internal, 5. T, 6. T

Chapter 5

Checking Geometry

INTRODUCTION

While meshing a model, Flow Simulation first interprets the specified solid and fluid regions as bodies and then creates a computational mesh for those bodies. By using the **Check Geometry** tool, you can view the resulting solid and fluid bodies. Also, you can locate the gap and holes by using the **Leak Tracking** tool. All these tools are discussed next.

CHECK GEOMETRY

CommandManager:	Flow Simulation > Check Geometry
SOLIDWORKS Menus:	Tools > Flow Simulation > Tools > Check Geometry
Toolbar:	Flow Simulation Tools > Check Geometry

This tool is used to detect the geometry related issues that result in poor mesh creation in flow simulation. To do so, choose the **Check Geometry** tool from the **Flow Simulation CommandManager**; the **Check Geometry** dialog box is displayed, refer to Figure 5-1. The options in this dialog box are discussed next.

State Rollout

In this rollout, you can select the component to test only or to apply the current project settings. The components of the assembly are displayed in the **Component** list of the **State** rollout. You can select the check box located on the right to the name of components to change the state of component. Choose the **Apply to Project** button to apply the settings to the project.

Note

*The **Apply to Project** button will be available only after the project has been created.*

Analysis Type Rollout

In this rollout, you can define an analysis type or select the check box for the problem you intend to solve with flow simulation. If the flow simulation project has not been created, you can choose between the **Internal** and **External** radio buttons from this rollout. The **Exclude cavities without flow conditions** check box can be selected to exclude closed internal spaces with no boundary condition but this check box will only be available when you have created a project. Select the **Exclude internal space** check box to ignore closed internal spaces in external flow analysis.

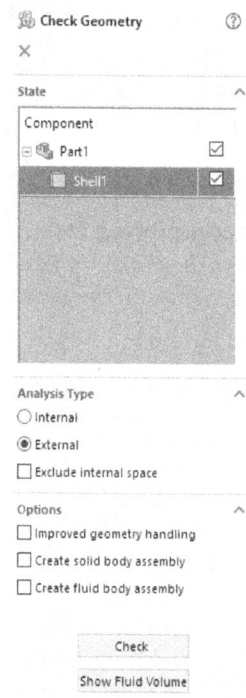

*Figure 5-1 The **Check Geometry** dialog box*

Options Rollout

The check boxes available in this rollout are explained next.

Improved geometry handling

Select this check box to use Flow Simulation algorithm to get solid and fluid bodies.

Create solid body assembly
Select this check box to create an assembly of solid bodies.

Create fluid body assembly
Select this check box to create an assembly of fluid bodies.

The **Check** button is used to calculate the total volume of solids and total fluid volume. The **Show Fluid Volume** button is used to show the fluid region in the drawing area.

LEAK TRACKING

CommandManager:	Flow Simulation > Tools > Leak Tracking
SOLIDWORKS Menus:	Tools > Flow Simulation > Tools > Leak Tracking
Toolbar:	Flow Simulation Tools Group > Leak Tracking

The **Leak Tracking** tool helps you to locate the gaps and holes in the model. Choose the **Leak Tracking** tool from the **Tools** drop-down of the **Flow Simulation CommandManager**; the **Leak Tracking** tab is added at the lower left corner of the drawing area, refer to Figure 5-2. In the **Start Face** area, you need to select the inner or outer face of the component and in the **End Face** area, you need to select the outer or inner face of the component.

*Figure 5-2 The **Leak Tracking** tab*

Choose the **Find Connection** button to detect a gap or hole. You can also interrupt the process by clicking the **Stop** button. If the hole is found, a path connecting the outside and inner faces will be highlighted in the drawing area. The list of faces connected by the path can be found in the **Track Faces** list, refer to Figure 5-3.

*Figure 5-3 The list of faces displayed in the **Track Faces** list*

TUTORIALS

To perform the tutorial, you need to download the zipped file named as *c05_SWF_2023_input* from the **Input Files** section of the CADCIM website. The complete path for downloading the file is:

Textbooks > CAE Simulation > Dassault Systemes > SOLIDWORKS Flow Simulation > Flow Simulation Using SOLIDWORKS 2023 > Input Files

After the file is downloaded, extract the folder to the location *C:\SOLIDWORKS_Flow\resources* and rename it as *c05*.

Tutorial 1

In this tutorial, you will open *c05_tut01_inp* file that you downloaded from the CADCIM website *www.cadcim.com*. You will use the tools to find any openings in the model. The model is shown in Figure 5-4.

(Expected time: 30 min)

The following steps are required to complete this tutorial:

a. Open the downloaded file.
b. Save this document in the c05 folder with a new name.
c. Suppress the computational domain.
d. Use the **Check Geometry** tool.
e. Use the **Leak Tracking** tool.
f. Delete the hole.
g. Use the **Check Geometry** tool.
h. Save the document.

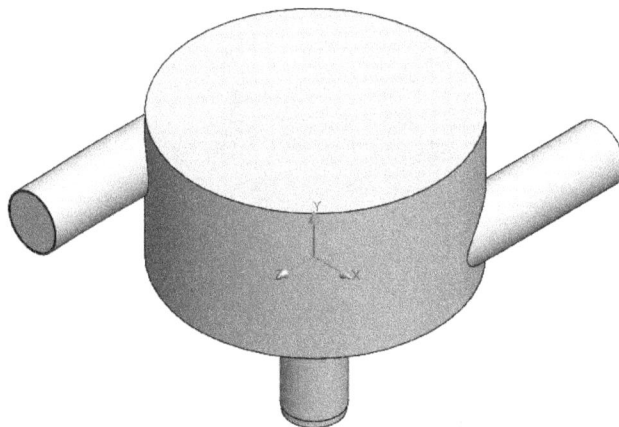

Figure 5-4 *Solid model for Tutorial 1*

Opening the Downloaded File

As the required document is saved in the *c05* folder, you need to select this folder and then open the *c05_tut01_inp.sldprt* document.

1. Start SOLIDWORKS by double-clicking on its shortcut icon on the desktop of your computer.

2. Choose the **Open** button from the Menu Bar to display the **Open** dialog box.

3. Browse to the SOLIDWORKS folder and select the **c05** folder.

4. Select the **c05_tut01_inp.sldprt** document and then choose the **Open** button; the model opens in Flow Simulation environment, refer to Figure 5-5.

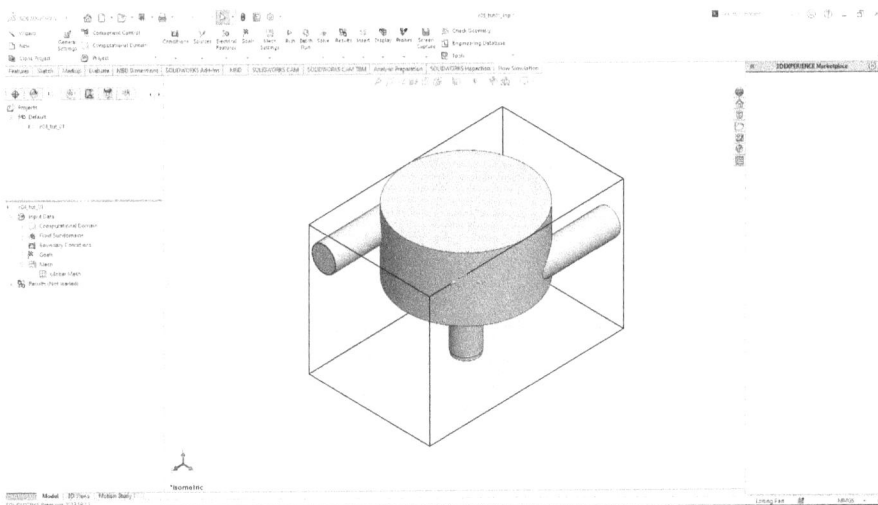

Figure 5-5 Solid model opened in Flow Simulation environment

Saving the Document in the c05 Folder

When you open a document from another chapter, it is recommended that you first save the opened document with a new name in the folder of the current chapter to avoid the original document from getting modified.

1. Choose the **Save As** button from the **Save** flyout in the Menu Bar; the **Save As** dialog box is displayed, refer to Figure 5-6.

2. Browse to the SOLIDWORKS folder and then create a new folder with the name **c05** by using the **Create New Folder** button. Make the **c05** folder as the current folder by double-clicking on it.

3. Enter **c05_tut01** as the new name of the document in the **File name** edit box and then choose the **Save** button to save the document.

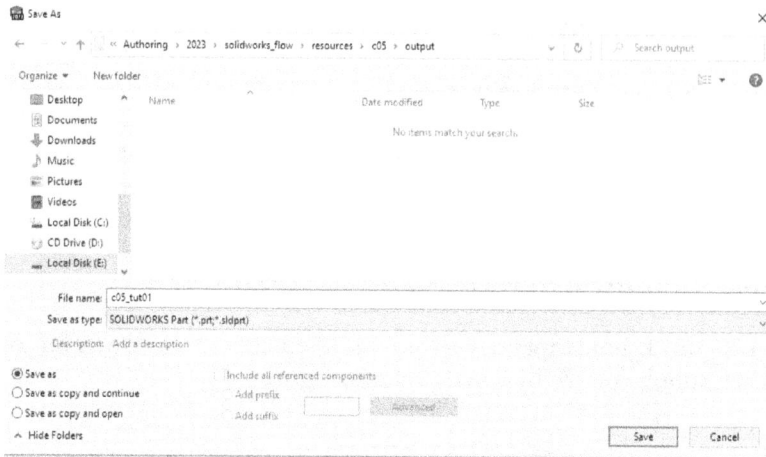

Figure 5-6 The Save As dialog box

4.   The document is saved with the new name and gets opened in the drawing area.

Hiding Computational Domain

Next, you need to suppress the computational domain from the flow simulation analysis tree.

1.   Select the **Computational Domain** from the **Flow Simulation Design** tree and right-click on it; a shortcut menu is displayed. Select the **Hide** option from the shortcut menu; the computational domain gets hidden in the graphics area, refer to Figure 5-7.

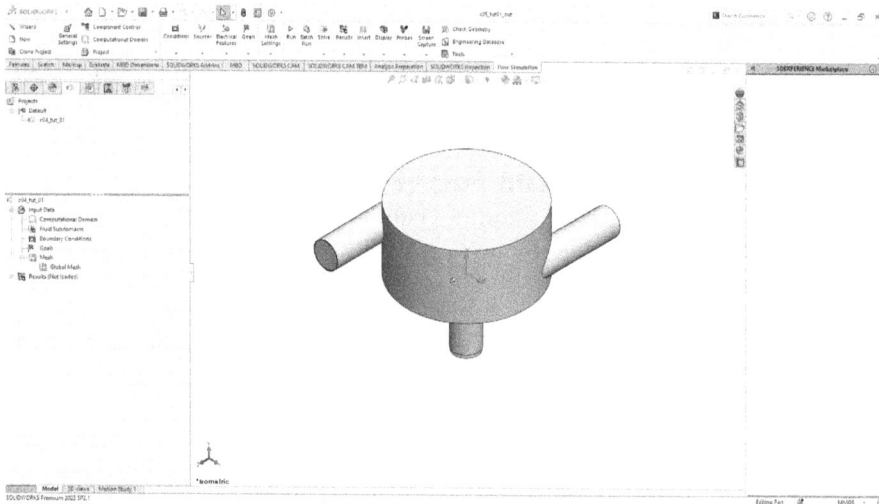

Figure 5-7 Solid model after suppressing computational domain

Checking the Geometry

Next, you need to use the **Check Geometry** tool to determine whether the geometry succeeded or failed.

1. Choose the **Check Geometry** tool from the **Tools CommandManager**; the **Check Geometry PropertyManager** is displayed, refer to Figure 5-8.

2. Select all the check boxes on the right of all the components in the **State** rollout so that the components are enabled for the test or you can apply the current project settings. If they are already selected then there is no need to select them.

3. Select the **Exclude cavities without conditions** check box, if not already selected, from the **Analysis Type** rollout to exclude closed internal spaces with no boundary condition specified on their surfaces.

4. Select the **Create solid body assembly** and **Create fluid body assembly** check boxes from the **Options** rollout to create an assembly of solid and fluid bodies.

5. Choose the **Check** button to operate fluid simulation with the model components to get solid and fluid bodies; the **Flow Simulation 2023** dialog box is displayed along with the **Check Geometry**

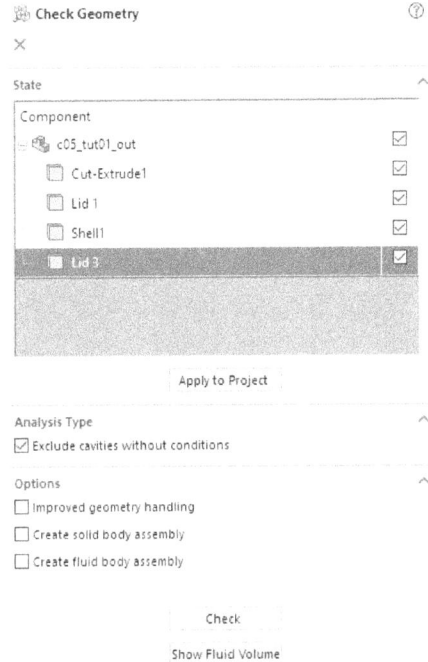

Figure 5-8 The *Check Geometry PropertyManager*

tab. This dialog box displays the following message: "**Cannot create a solid body. Solid volume is zero. Cannot create fluid body. Fluid volume is zero**".

6. Choose the **OK** button to close the dialog box. You will notice that the **Status is failed** and **Non watertight model** are displayed in the **Results** list of the **Check Geometry** tab, refer to Figure 5-9. As the model does not have any closed internal volumes, so you must close openings and holes to make the internal volume closed.

Figure 5-9 The *Results* list

7. Choose **Close** and then the **Cancel** button to close the **Check Geometry** tab and the **Check Geometry PropertyManager**, respectively.

Tracking Leaks in the Model

Next, you will check the opening by using the **Leak Tracking** tool.

1. Choose the **Leak Tracking** tool from the **Tools** drop-down of the **Flow Simulation Command Manager**; the **Leak Tracking** tab is added at the lower left corner of the drawing area, refer to Figure 5-10.

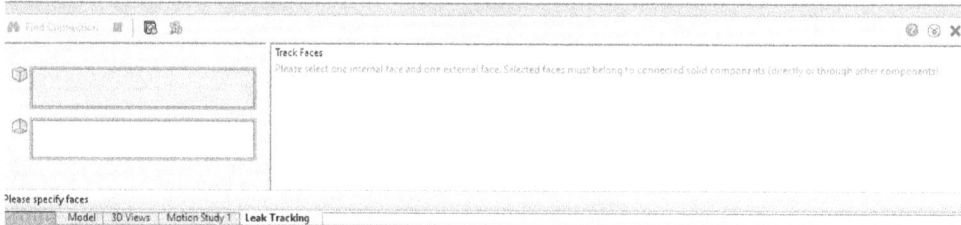

Figure 5-10 The Leak Tracking tab

Next, you need to select the inner and outer faces of the lid. To select the faces, you need to section the model by using the **Section View** tool.

2. Choose the **Section View** tool from the **View (Heads-Up)** toolbar; the **Section View PropertyManager** is displayed.

3. Create the section view of the model, as shown in Figure 5-11 and choose the **OK** button to close the PropertyManager.

4. Select the inside face of the Lid 1 in the **Start Face** area of the **Leak Tracking** tab. Refer to Figure 5-12 for face selection.

5. Select the outside face of the Lid 3 in the **End Face** area of the **Leak Tracking** tab. Refer to Figure 5-13 for face selection.

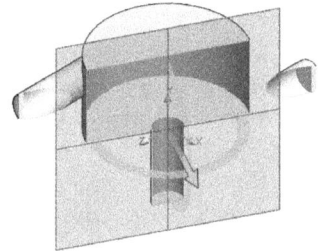

Figure 5-11 The section view of the model

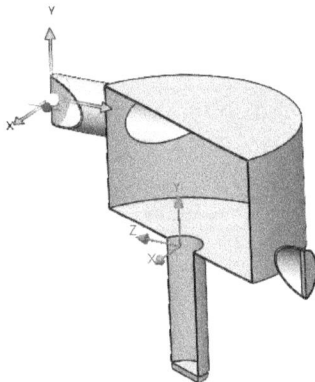

Figure 5-12 The Lid 1 face selected

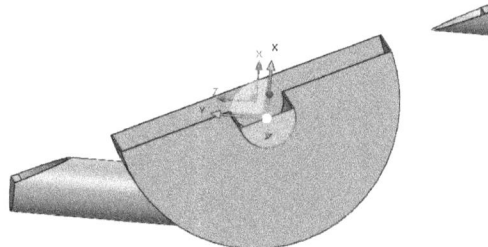

Figure 5-13 The Lid 3 face selected

6. Choose the **Find Connection** button from the **Leak Tracking** tab; the checking process start and you will see the list of faces connected by the highlighted path in the **Track Faces** list, refer to Figure 5-14.

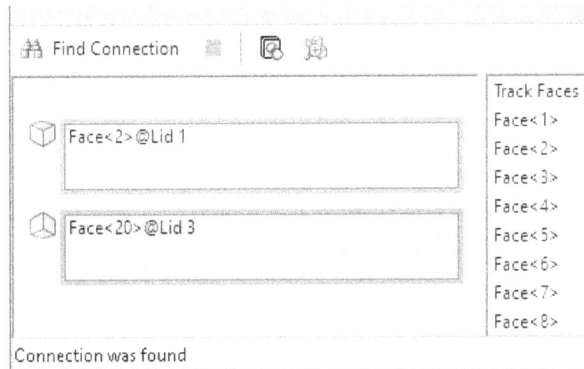

Figure 5-14 *The face listed in* **Track Faces** *list*

7. Select the detected faces from the list to highlight the faces along the path that connects the inner and outer faces one by one.

8. Select **Face<6>** from the **Track Faces** list. You will see that the highlighted area in the drawing area has a hole in the Lid 2. This hole needs to be deleted.

Deleting the Hole

Next, you need to delete the hole that was found during the leak detection process. Before deleting the hole, you need to restore the view of the model.

1. Choose the **Section View** tool from the **View (Heads-Up)** toolbar; the view of the model is restored, refer to Figure 5-15.

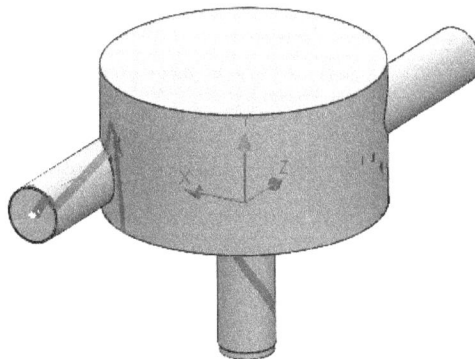

Figure 5-15 *The restore view of the model*

2. Click on the **FeatureManager Design Tree** and then select the hole which is detected in the previous step; it is highlighted in the **FeatureManager Design Tree**.

3.   Next, right-click to invoke the shortcut menu and choose the **Delete** option; the **Confirm Delete** dialog box is displayed, refer to Figure 5-16. Select the **Delete absorbed features** check box.

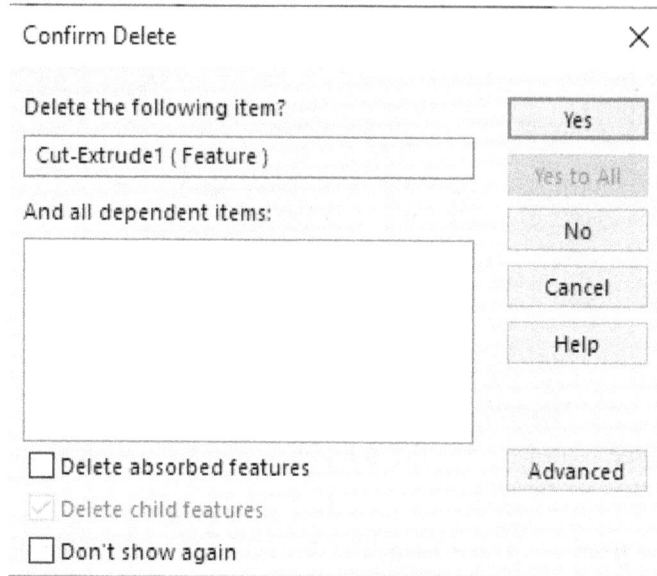

Figure 5-16 The Confirm Delete dialog box

4.   Choose the **Yes** button to close the dialog box; the **Flow Simulation 2023** dialog box is displayed.

5.   Choose the **Yes** button twice as the **Flow Simulation 2023** dialog box is displayed two times; the hole is deleted from the model.

Checking the Geometry

Next, you need to use the **Check Geometry** tool to determine whether the geometry succeeded or failed. Before using this tool, you need to use the **Section View** tool to select the faces.

1.   Choose the **Section View** tool from the **View (Heads-Up)** toolbar; the section view of the model is created, refer to Figure 5-17.

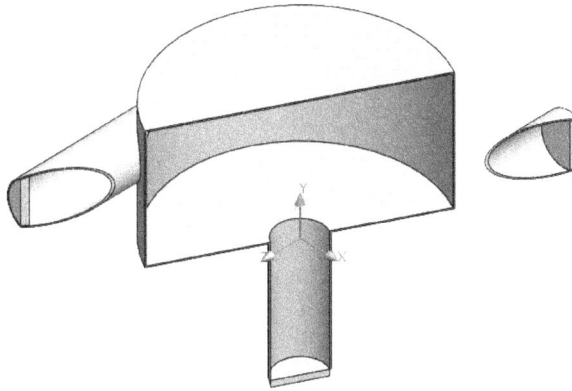

Figure 5-17 *Section view of model*

2. Choose the **Check Geometry** tool from the **Flow Simulation CommandManager**; the **Check Geometry PropertyManager** is displayed.

3. Select all the check boxes, if they are not selected, to the right of all the components in the **State** rollout so that the components are enabled for the test or you can apply the current project settings.

4. Clear the **Exclude cavities without conditions** check box from the **Analysis Type** rollout.

5. Select the **Create solid body assembly** and **Create fluid body assembly** check boxes from the **Options** rollout to create an assembly of the solid and fluid bodies.

6. Choose the **Check** button to operate fluid simulation with the model components to get solid and fluid bodies; the **Check Geometry** tab is displayed along with the two part files having name Part1 and Part2, refer to Figure 5-18 and 5-19. These part files contain solid body and fluid body. You will notice that the status is SUCCESSFUL.

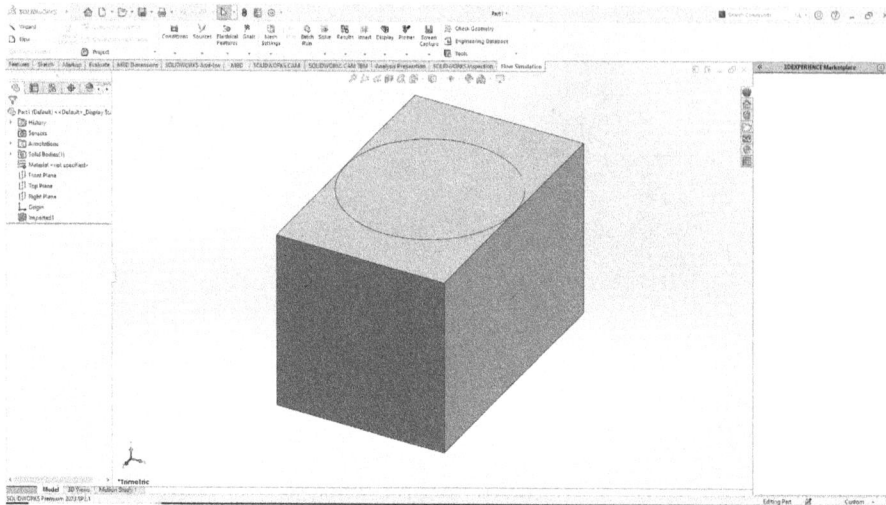

Figure 5-18 *The Part1 file*

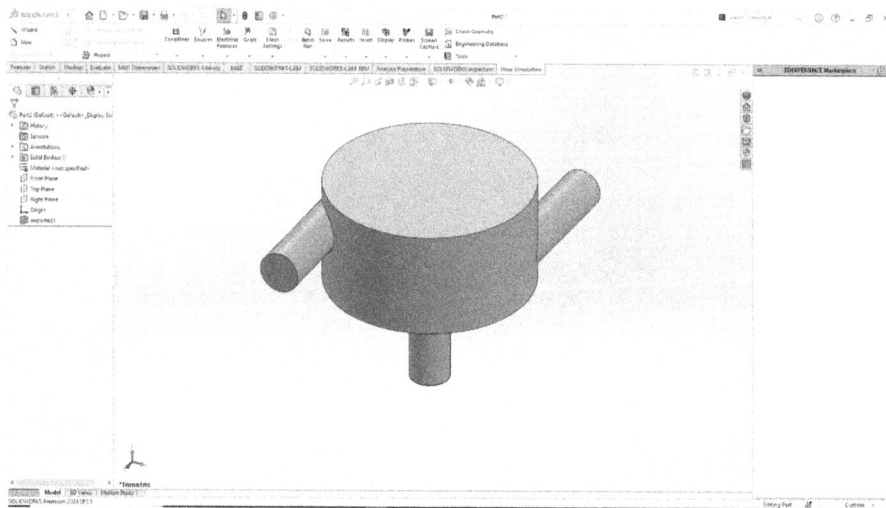

Figure 5-19 *The Part2 file*

7. Choose the **Close** and then the **Cancel** button to close the **Check Geometry** tab and **Check Geometry PropertyManager**.

Saving the Model

1. Save the part document with the name **c05_tut01** at the following location: *SOLIDWORKS_ Flow\resources\c05*.

2. Choose **File > Close** from the SOLIDWORKS menus to close the document.

Self-Evaluation Test

Answer the following questions and then compare them to those given at the end of this chapter:

1. Which of the following PropertyManagers is displayed when you choose the **Check Geometry** tool from the **Flow Simulation CommandManager**?

   (a) **Geometry**   (b) **Check Geometries**
   (c) **Check Geometry**   (d) None of these

2. Which of the following buttons is used to show the fluid region in the drawing area?

   (a) **Show Volume**   (b) **Show Fluid Volume**
   (c) **Fluid**   (d) None of these

3. The _____ tool is used to check the geometry issues in SOLIDWORKS Flow Simulation.

4. The _____ tool is used to locate holes in a model.

EXERCISES
Exercise 1

In this exercise, you will open the model (*c05_exr01*) created in Exercise 1 of Chapter 5. You can also download this file from *www.cadcim.com* by using the following path:

Textbooks > CAE Simulation > Dassault Systemes > SOLIDWORKS Flow Simulation > Flow Simulation Using SOLIDWORKS 2023 > Input Files > C05_SWFS_inp

You will then use the **Check Geometry** tool to validate the geometry. The model is shown in Figure 5-20.

(**Expected time: 10 min**)

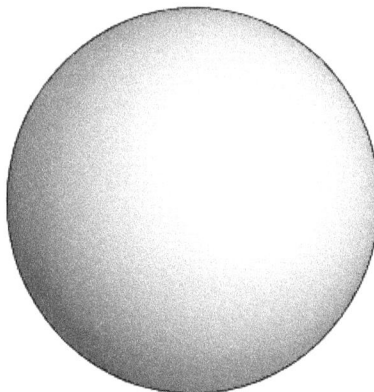

Figure 5-20 *Solid model for Exercise 1*

Exercise 2

In this exercise, you will open the model (*c05_exr02*) created in Exercise 2 of Chapter 5. You can also download this file from *www.cadcim.com* by using the following path:

Textbooks > CAE Simulation > Dassault Systemes > SOLIDWORKS Flow Simulation > Flow Simulation Using SOLIDWORKS 2023 > Input Files > C05_SWFS_inp

You will then use the **Check Geometry** tool to validate the geometry. The model is shown in Figure 5-21.

(Expected time: 10 min)

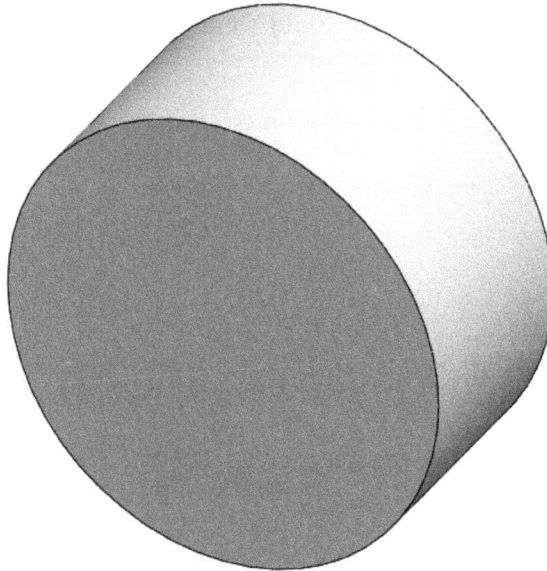

Figure 5-21 Solid model for Exercise 2

Answers to Self-Evaluation Test
1. c, 2. b, 3. Check Geometry, 4. Leak Tracking

Chapter 6

Boundary Conditions

Learning Objectives

After completing this chapter, you will be able to:
- *Define and edit a computational domain*
- *Apply boundary conditions*
- *Define a surface source*

INTRODUCTION

To define a problem that would result in a unique solution, you need to specify variables at the boundaries of the model. These variables indicate the limitations or constraints and are called as boundary conditions. You need to identify the location and apply input parameters at the boundary. You will learn about it later in the chapter.

COMPUTATIONAL DOMAIN

CommandManager:	Flow Simulation > Computational Domain
SOLIDWORKS Menus:	Tools > Flow Simulation > Computational Domain
Toolbar:	Flow Simulation Main > Computational Domain

The computational domain is the area where flow computations take place. When you create a new project, a computational domain enclosing the model is automatically created. In 2D and 3D simulations, a rectangular parallelepiped is used as the computational domain.

After specifying the project settings, you will notice that an input data folder is available in the **FeatureManager Design Tree**. Under this input data folder, a computation domain is displayed. When you click on this computation domain, you will notice that a box enclosing the model is displayed. Also, arrows appear on the edge of the box. These arrows can be used to change the size of the computational domain.

To create a computational area, choose the **Computational Domain** tool from the **Flow Simulation CommandManager**; the **Computational Domain PropertyManager** is displayed, as shown in Figure 6-1. By default, the **3D simulation** button is chosen in the **Type** rollout of the **Computational Domain PropertyManager**. You can edit the size of the computational domain by specifying the values in the **Size and Conditions** rollout. There are six edit boxes under the **Size and Conditions** rollout. The **X max** and **X min** edit boxes are used to change the size of the box in the positive and negative X directions, respectively. In the same way, you can specify the value for the Y and Z directions. Choose the **Reset** button to reset the values in the edit box in the **Size and Conditions** rollout. Click on the **Edge Color** and **Face Color** down arrows in

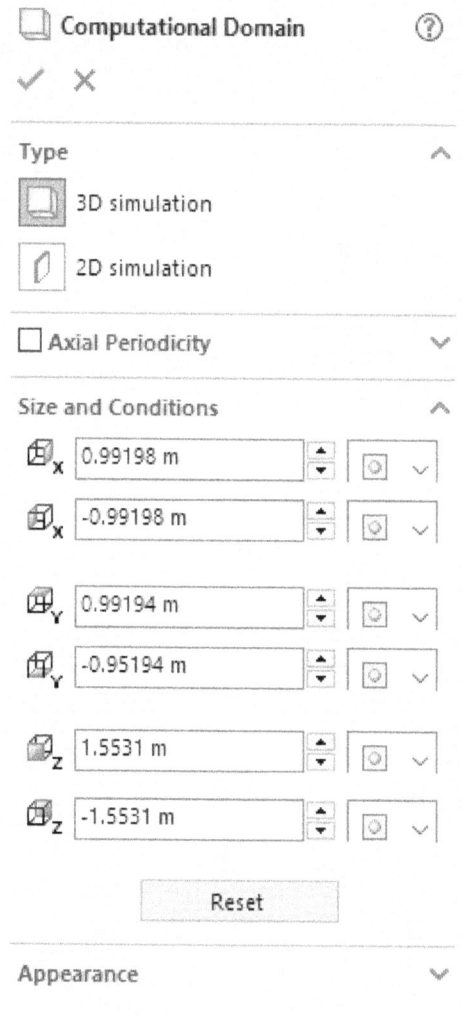

Figure 6-1 The Computational Domain PropertyManager

the **Appearance** rollout to change the appearance of the computational domain frame. You can also adjust the transparency of all computational domain faces with the Face Transparency slider.

Choose the **2D simulation** button from the **Type** rollout to simulate the 2D flow. This simulation helps to reduce the computation time. As you choose this button, the **XY plane**, **YZ plane**, and **XZ plane** radio buttons will be available in the **Type** rollout. As you select the **XY plane** radio button, the computational domain frame is created symmetrically about the XY plane. Similarly, symmetry will be created when the **XZ** and **YZ plane** radio buttons are selected.

BOUNDARY CONDITIONS

CommandManager: Flow Simulation > Conditions > Boundary Condition
SOLIDWORKS menus: Tools > Flow Simulation > Insert > Boundary Condition
Toolbar: Flow Simulation Features Group > Computational Domain

Flow Simulation requires boundary conditions to solve a problem. A steady-state flow pattern depends entirely on boundary conditions, whereas a time-dependent flow pattern depends on both boundary conditions and initial conditions.

To apply the boundary condition, choose the **Boundary Condition** tool from the **Conditions** drop-down in the **Flow Simulation CommandManager**; the **Boundary Condition Property Manager** is displayed, refer to Figure 6-2. The rollouts available in this dialog box are discussed next.

Selection
In this rollout, the faces are selected in the **Faces to Apply the Boundary Condition** selection box and the boundary conditions are applied to those faces. If you want to delete a particular face from the list of faces then select it and press the delete key from the keyboard. Choose the **Filter Faces** button if you want to remove a face of a specified type from the list of selected faces. As you choose this button, the **Filter Faces** rollout becomes available in the Property Manager which is discussed next.

Filter Faces
In this rollout, four options are available. These options are explained next.

Remove out of domain faces
This option is used when the faces lying outside the computational domain need to be removed from the faces list.

Figure 6-2 The Boundary Condition PropertyManager

Remove outer faces
This is used only for internal analysis. When this option is selected, the outer faces of the model are removed from the faces list.

Remove fluid-contacting faces
When this option is selected, the faces lying on the solid/fluid interface are removed from the faces list. The solid/porous and fluid/porous interfaces remain on the list.

Keep outer and fluid-contacting faces
This option allows the face list to contain only the faces at the solid/fluid interface (including the model's outer faces) for internal analysis. Selecting this option will remove from the face list all faces lying at the solid/solid, solid/porous, and porous/fluid interfaces, as well as faces outside the computational domain.

Type
In this rollout, three buttons are available and they are discussed next.

Flow Openings
This button is used when parameters such as mass flow, mass flux, volume flow, and velocity are required as boundary conditions. If you choose the **Flow Openings** button from the **Type** rollout, the **Flow Parameters** rollout will be displayed. This rollout will be discussed later in this chapter. In the Flow Openings, you can select the following parameters:

a) Inlet Mass Flow
b) Inlet Mass Flux
c) Inlet Volume Flow
d) Inlet Velocity
e) Outlet Mass Flow
f) Outlet Volume Flow
g) Outlet Velocity

Pressure Openings
This button is used when parameters such as Environment Pressure, Static Pressure, and Total Pressure are required as boundary conditions. The parameters are discussed next.

Environment Pressure: It is impossible to measure environmental pressure in the real world because it is not physical pressure. Flow Simulation provides a boundary condition option in its virtual environment. Simulations interpret this pressure as a Static Pressure for outlet flow and as a Total Pressure for the inlet flow.

Static Pressure: The static pressure is the pressure at a single point in a flow field that changes continuously with the flow of fluid or gas. You can check the pressure of a liquid without making it go faster or slower when it goes over a thing that stays still or moves with the liquid's movement.

Total Pressure: The pressure of a fluid flow when it is stationary is known as total pressure. It is also called Stagnation Pressure.

Wall

This button is used to specify the wall conditions on the selected face.

Real Wall: You can specify roughness, heat transfer coefficents, and wall temperatures for chosen fluid-contacting walls by using this condition. The Real Wall condition also allows you to specify tangential velocity boundary conditions at a wall to simulate translation or rotation.

Ideal Wall: The Ideal Wall condition allows you to specify adiabatic, frictionless walls for all selected faces instead of default fluid friction walls.

Flow Parameters

In this rollout, you can specify speed, Mach number (only for gases), mass flow rate, and volume flow rate. There are three buttons available in this rollout. They are discussed next.

Normal To Face

This button is used when the flow direction is perpendicular to the aperture surface.

Swirl

This button is used when you need to specify the swirling of the flow about an axis of the reference Coordinate System.

3D Vector

This button is used when you need to specify the flow velocity or direction vector by its components in X-, Y- and Z-directions.

Thermodynamic Parameters

The options in this rollout are used to set the pressure and temperature for the pressure openings and inlet flow openings.

Turbulence Parameters

The options in this rollout are used to specify parameters for turbulence in the inlet flow.

Boundary Layer

The options in this rollout are used to specify the parameters of boundary layer in the inlet flow.

Goals

The rollout has options to set goals to associate with the boundary condition.

Name Template

In the Name Template edit box in the **Name Template** rollout, you will notice <Type> and <Number>. They represent the boundary condition type and the sequential number of the boundary condition, respectively in the flow simulation analysis.

Choose the **OK** button after specifying the boundary conditions to close the **Boundary Condition PropertyManager**.

CREATING A SURFACE SOURCE

CommandManager:	Flow Simulation > Sources > Surface Source
SOLIDWORKS menus:	Tools > Flow Simulation > Insert > Surface Source
Toolbar:	Flow Simulation Features Group > Surface Source

The **Surface Source** tool allows you to define a heat surface source on a surface that is in touch with fluid as well as a surface that is a solid-solid boundary.

To apply the surface source on a surface, choose the **Surface Source** tool from the **Conditions** drop-down in the **Flow Simulation CommandManager**; the **Surface Source Property Manager** is displayed, refer to Figure 6-3. The rollouts available in this dialog box are discussed next.

Selection

Select the faces on which you want to apply the surface source. The selected faces are displayed in the **Faces to Apply the Surface Source** area under the **Selection** rollout. If you want to delete a particular face from the list of faces then select that face and press the delete key from the keyboard. Click on the **Filter Faces** button if you want to remove unnecessary faces of a specified type from the list of selected faces. As you select this button, the **Filter Faces** rollout becomes available in the Property Manager which will be discussed next.

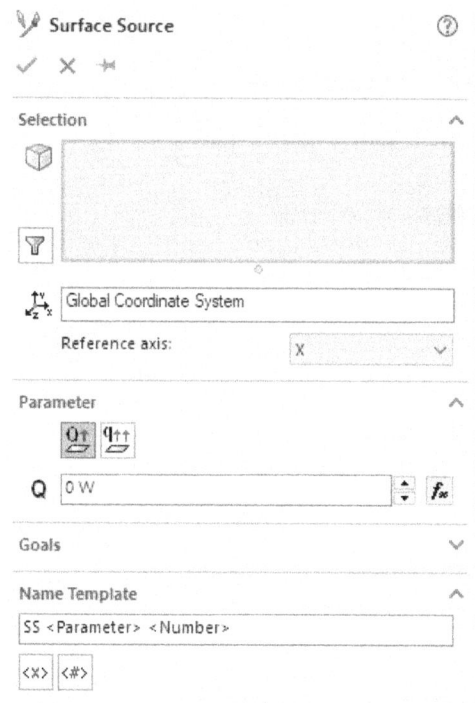

Figure 6-3 The Surface Source PropertyManager

Remove out of domain faces
This option is used when the faces lying outside the computational domain need to be removed from the faces list.

Remove outer faces
This is used only for internal analysis. When this option is selected, the model outer faces are removed from the faces list.

Remove fluid-contacting faces
When this option is selected, the faces lying on the solid/fluid interface are removed from the faces list. The solid/porous and fluid/porous interfaces remain on the list.

Keep outer and fluid-contacting faces

This option allows the face list to contain only the faces at the solid/fluid interface (including the model's outer faces for internal analysis). Selecting this option will remove from the face list all faces lying at the solid/solid, solid/porous, and porous/ fluid interfaces, as well as faces outside the computational domain.

Parameter

In this rollout, you are provided with the options for specifying the heat transfer rate and the heat flux. These options are discussed next.

Heat Transfer Rate

The heat transfer rate between two surfaces is the ratio of temperature difference and the thermal resistance between them. When you choose this button, you can specify its value in the edit box available in this rollout.

Heat Flux

The heat flux is a flow of energy per unit of area per unit of time. When you choose this button, you can specify its value in the edit box available in this rollout.

Choose the **OK** button after specifying the parameters to close the **Surface Source PropertyManager**.

TUTORIALS

Tutorial 1

In this tutorial, you will open the model (*c04_tut01*) created in Tutorial 1 of Chapter 4. You can also download this file from *www.cadcim.com* by using the following path:

Textbooks > CAE Simulation > Dassault Systemes > SOLIDWORKS Flow Simulation > Flow Simulation Using SOLIDWORKS 2023 > Input Files > C06_SWFS_inp

After opening the file, you will then apply the boundary conditions to the model. The model is shown in Figure 6-4.   **(Expected time: 20 min)**

The following steps are required to complete this tutorial:

a. Open Tutorial 1 of Chapter 4.
b. Save this tutorial in the *c06* folder with a new name.
c. Add boundary conditions to the model.
d. Save the file.

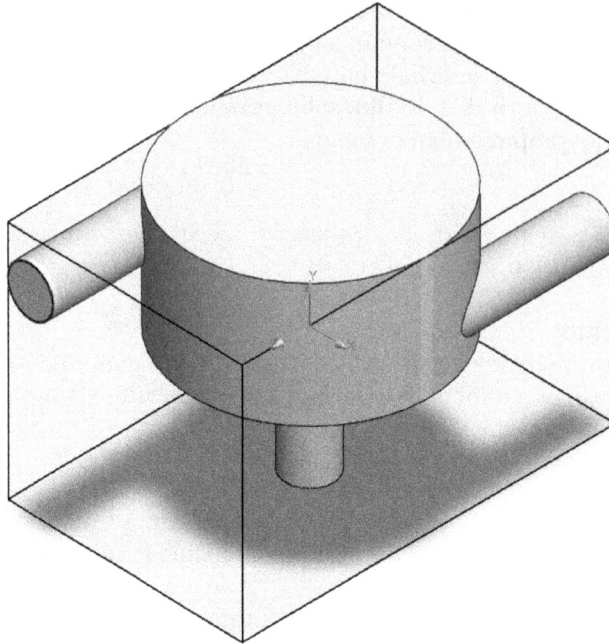

Figure 6-4 *Model for Tutorial 1*

Opening Tutorial 1 of Chapter 4

As the required tutorial is saved in the *c04* folder, you need to select this folder and then open the *c04_tut01.sldprt* document.

1. Start SOLIDWORKS by double-clicking on its shortcut icon on the desktop of your computer.

2. Choose the **Open** button from the Menu Bar to display the **Open** dialog box, refer to Figure 6-5.

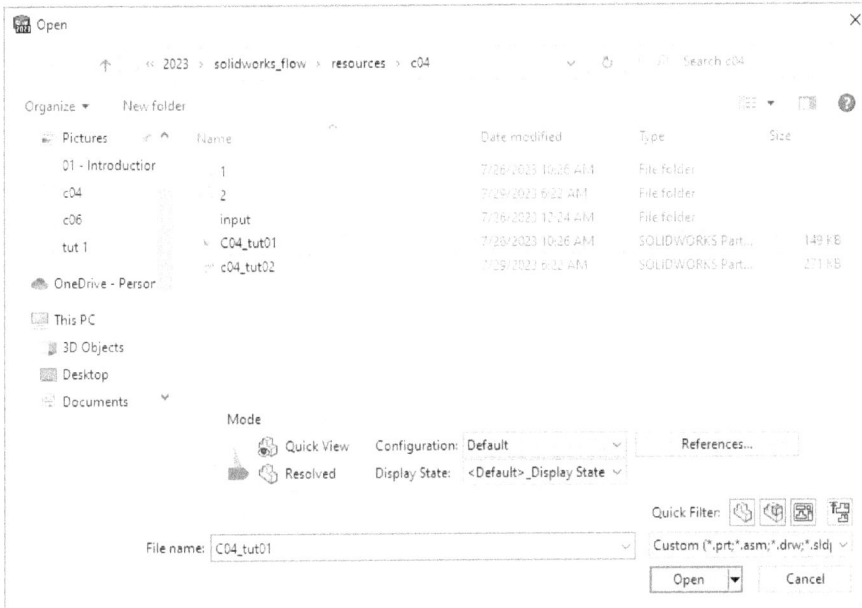

Figure 6-5 *The **Open** dialog box*

3. Browse to the SOLIDWORKS folder and select the **c04** folder.

4. Select the **c04_tut01.sldprt** document and then choose the **Open** button.

As the model was saved in the Flow Simulation environment in Chapter 4, it opens in the Flow Simulation environment, refer to Figure 6-6.

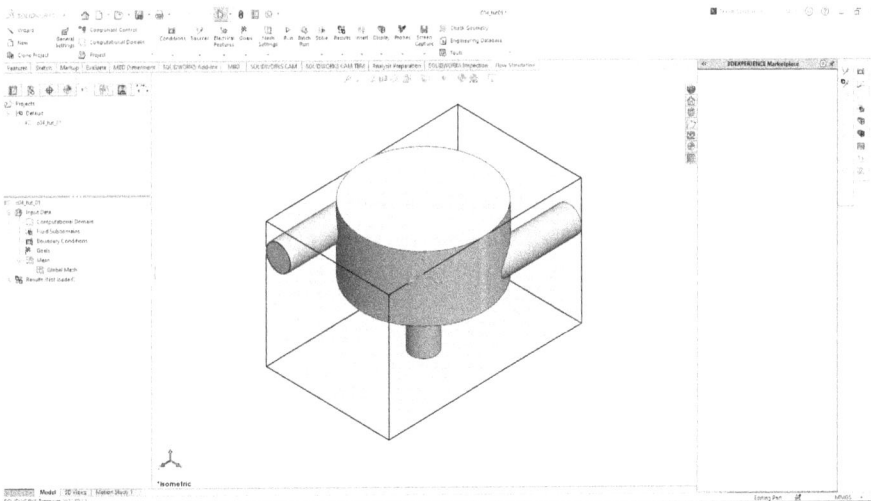

Figure 6-6 *The model opened in the Flow Simulation environment*

Saving the Document in the c06 Folder

When you open a document from another chapter, it is recommended that you first save the opened document with a new name in the folder of the current chapter to avoid the original document from getting modified.

1. Choose the **Save As** button from the **Save** flyout in the Menu Bar; the **Save As** dialog box is displayed, refer to Figure 6-7.

2. Browse to the **SOLIDWORKS_Flow > Resources** folder and then create a new folder with the name **c06** by using the **Create New Folder** button. Make the **c06** folder as the current folder by double-clicking on it.

3. Enter **c06_tut01** as the new name of the document in the **File name** edit box and then choose the **Save** button to save the document.

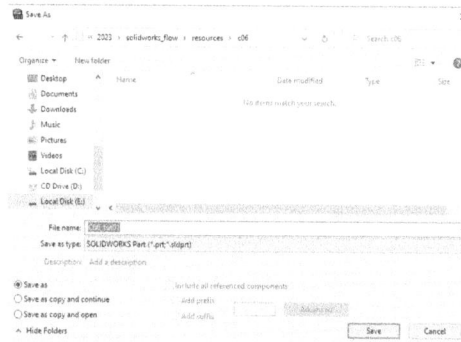

Figure 6-7 The Save As dialog box

4. The document is saved with the new name and gets opened in the drawing area.

Hiding the Computational Domain

Next, you need to hide the computational domain from the Flow Simulation Analysis tree.

1. Select the **Computational Domain** from the **Flow Simulation Design** tree and right-click on it; a shortcut menu is displayed. Choose the **Hide** option from the shortcut menu; the computational domain gets hidden in the graphics area, refer to Figure 6-8.

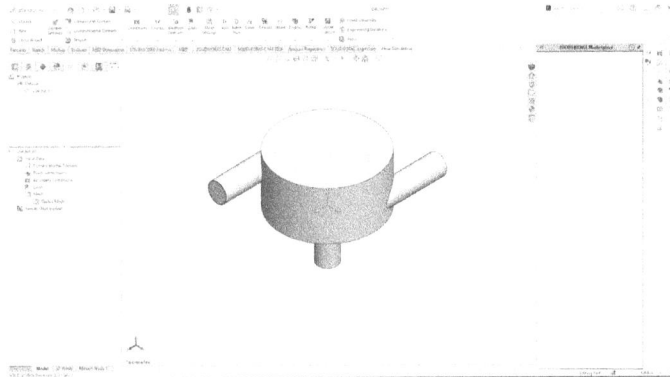

Figure 6-8 *The model without computational domain*

Creating the Section View

1. Choose the **Section View** tool from the **View (Heads-Up)** toolbar; the **Section View PropertyManager** is displayed.

2. Create the section view of the model, as shown in Figure 6-9 and choose the **OK** button to close the PropertyManager.

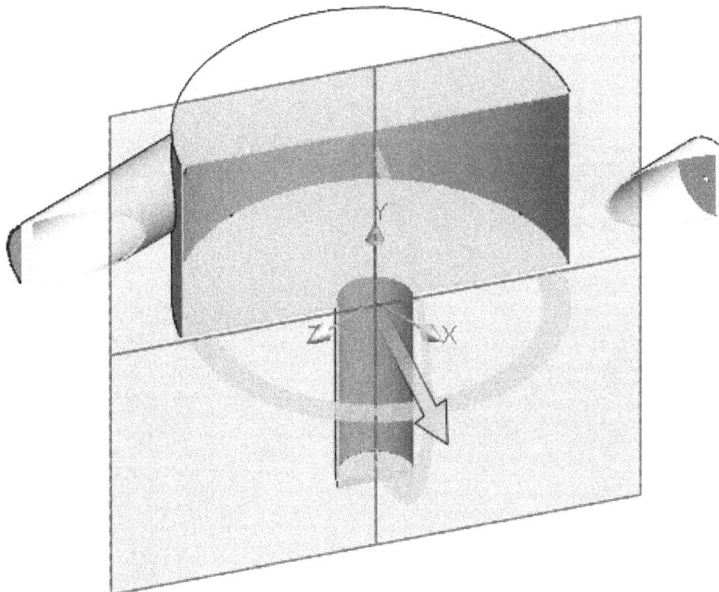

Figure 6-9 *The section view of the model*

Applying Boundary Condition to Lid 1

Next, you need to invoke the **Boundary Condition** tool and apply the boundary condition to the model.

1. Right-click on the **Boundary Conditions** in the **Flow Simulation Analysis** tree; a shortcut menu is displayed. Select the **Insert Boundary Condition** option from it; the **Boundary Condition PropertyManager** is displayed, refer to Figure 6-10.

2. Select the inside face of Lid1 from the graphics area, refer to Figure 6-11. The selected face is displayed in the **Faces to Apply the Boundary Condition** selection box in the **Selection** rollout.

3. Select **Inlet Velocity** from the **Type** rollout. Make sure that the **Flow Openings** button is selected in the **Type** rollout while selecting **Inlet Velocity**.

4. Enter **5** in the **Velocity Normal to Face** edit box and ensure that the **Normal to Face** button is selected in the **Flow Parameters** rollout.

5. Choose the **OK** button from the **Boundary Condition PropertyManager** to close it.

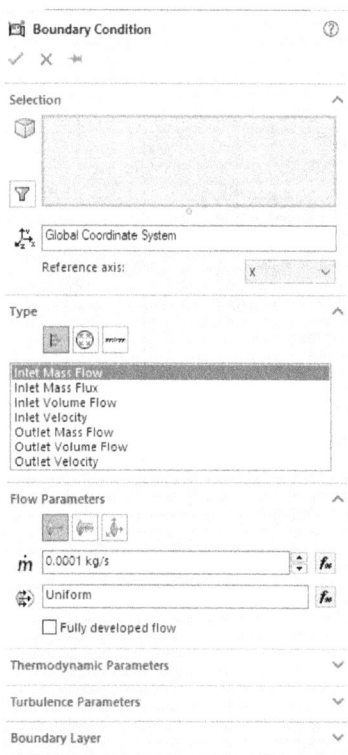

Figure 6-10 *The Boundary Condition PropertyManager*

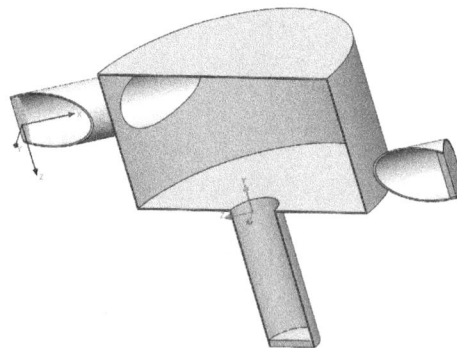

Figure 6-11 *The Lid 1 selected*

Applying Boundary Condition to Lid 2

Next, you need to apply the inlet velocity parameter to Lid 2.

1. Right click on the **Boundary Conditions** in the **Flow Simulation Analysis** tree; a shortcut menu is displayed. Select the **Insert Boundary Condition** option from it; the **Boundary Condition PropertyManager** is displayed

2. Select the inside face of Lid 2 from the graphics area, refer to Figure 6-12. The selected face is displayed in the **Faces to Apply the Boundary Condition** selection box in the **Selection** rollout.

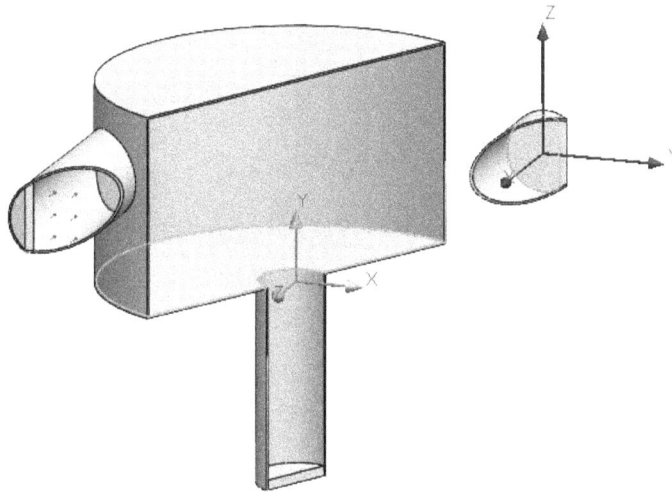

Figure 6-12 The Lid 2 selected

3. Select **Inlet Velocity** from the **Type** rollout. Ensure that the **Flow Openings** button is selected in the **Type** rollout while selecting **Inlet Velocity**.

4. Enter **3** in the **Velocity Normal to Face** edit box and ensure that the **Normal to Face** button is selected in the **Flow Parameters** rollout.

5. Choose the **OK** button from the **Boundary Condition PropertyManager** to close it.

Applying Boundary Condition to Lid 3

Next, you need to apply the pressure parameter to Lid 3.

1. Right-click on the **Boundary Conditions** in the **Flow Simulation Analysis** tree; a shortcut menu is displayed. Select the **Insert Boundary Condition** option from it; the **Boundary Condition PropertyManager** is displayed.

2. Select the inside face of Lid 3 from the graphics area, refer to Figure 6-13; the selected face is displayed in the **Faces to Apply the Boundary Condition** selection box in the **Selection** rollout.

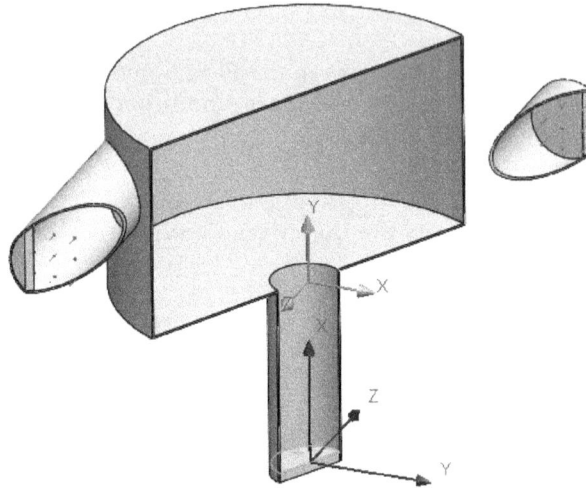

Figure 6-13 *The Lid 3 selected*

3. Select the **Static Pressure** option from the **Type** rollout. Ensure that the **Pressure Openings** button is chosen in the **Type** rollout while selecting the **Static Pressure** option.

4. Choose the **OK** button from the **Boundary Condition PropertyManager** to close it.

Turn off the Section View
Next, you need to turn off the section view of model.

1. Choose the **Section View** tool from the **View (Heads-Up)** toolbar; the full view of the model is displayed, refer to Figure 6-14.

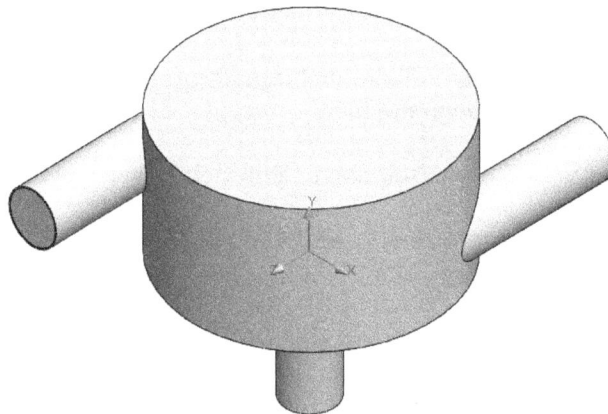

Figure 6-14 *The model after section off*

Saving the Model

1. Save the part document with the name *c06_tut01* at the following location: *SOLIDWORKS_Flow\resources\c06*.

2. Choose **File > Close** from the SOLIDWORKS menus to close the document.

Tutorial 2

In this tutorial, you will open the model (*c04_tut02*) created in Tutorial 2 of Chapter 4. You can also download this file from *www.cadcim.com* by using the following path:

Textbooks > CAE Simulation > Dassault Systemes > SOLIDWORKS Flow Simulation > Flow Simulation Using SOLIDWORKS 2023 > Input Files > C06_SWFS_inp

After opening the file, you will then apply the boundary conditions to the model. The model is shown in Figure 6-15.   **(Expected time: 20 min)**

The following steps are required to complete this tutorial:

a. Open Tutorial 2 of Chapter 4.
b. Save this tutorial in the *c06* folder with a new name.
c. Add boundary conditions to the model.
d. Save the file.

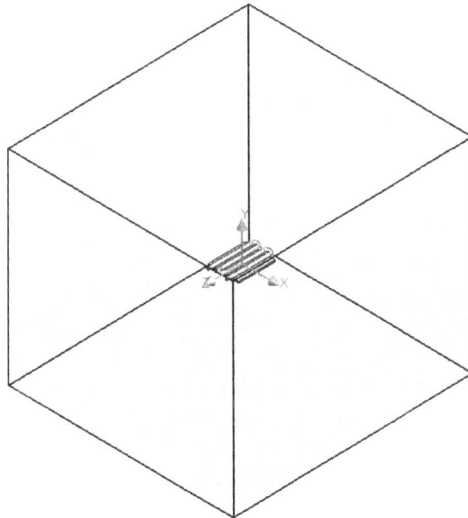

Figure 6-15 Model for Tutorial 1

Opening Tutorial 2 of Chapter 4

As the required document is saved in the *c04* folder, you need to select this folder and then open the *c04_tut02.sldprt* document.

1. Start SOLIDWORKS by double-clicking on its shortcut icon on the desktop of your computer.

2. Choose the **Open** button from the Menu Bar to display the **Open** dialog box.

3. Browse to the SOLIDWORKS folder and select the **c04** folder.

4. Select the **c04_tut02.sldprt** document and then choose the **Open** button.

   As the model was saved in the part modeling environment in Chapter 3, it opens in the part modeling environment.

Saving the Document

When you open a document from another chapter, it is recommended that you first save the opened document with a new name in the folder of the current chapter to avoid the original document from getting modified.

1. Choose the **Save As** button from the **Save** flyout in the Menu Bar; the **Save As** dialog box is displayed, as shown in Figure 6-16.

2. Browse to the **SOLIDWORKS_Flow > resources** folder and make the **c06** folder as the current folder by double-clicking on it.

3. Enter **c06_tut02** as the new name of the document in the **File name** edit box and then choose the **Save** button to save the document.

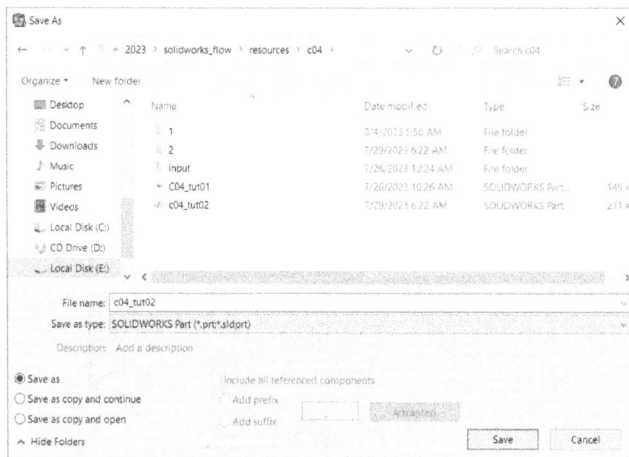

*Figure 6-16 The **Save As** dialog box*

4. The document is saved with the new name and gets opened in the drawing area.

Editing the Computational Domain

Next, you need to edit the computational domain from the flow simulation analysis tree.

1. Select the **Computational Domain** from the **Flow Simulation Design** tree and right-click on it; a shortcut menu is displayed. Choose the **Edit Definition** option from the shortcut menu; the **Computational Domain PropertyManager** is displayed, as shown in Figure 6-17.

2. Enter the following values in the **Size and Conditions** rollout, as shown in Figure 6-17.

3. Choose the **OK** button to close the PropertyManager.

    After specifying the values, the computational domain size will look like shown in Figure 6-18.

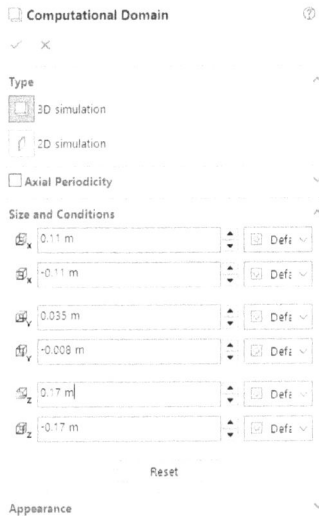

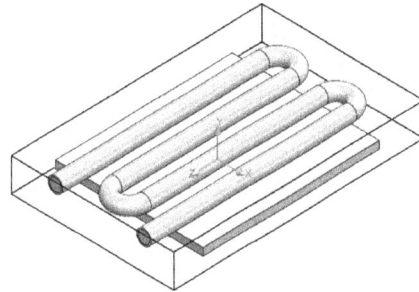

Figure 6-17 The **Computational Domain** *dialog box*   *Figure 6-18* *The computational domain boundary*

Hiding the Computational Domain

Next, you need to suppress the computational domain from the flow simulation analysis tree.

1. Select the **Computational Domain** from the **Flow Simulation Design** tree and right-click on it; a shortcut menu is displayed. Choose the **Hide** option from the shortcut menu, the computational domain gets hidden in the graphics area. Refer to Figure 6-19 after hiding the computational domain.

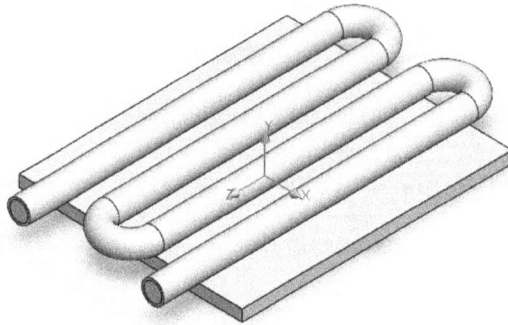

Figure 6-19 *The model without computational domain boundary*

Section View of the Model

Next, you need to create the section view of the model.

1. Choose the **Section View** tool from the **View (Heads-Up)** toolbar; the **Section View PropertyManager** is displayed.

2. Create the section view of the model as shown in Figure 6-20 and choose the **OK** button to close the PropertyManager.

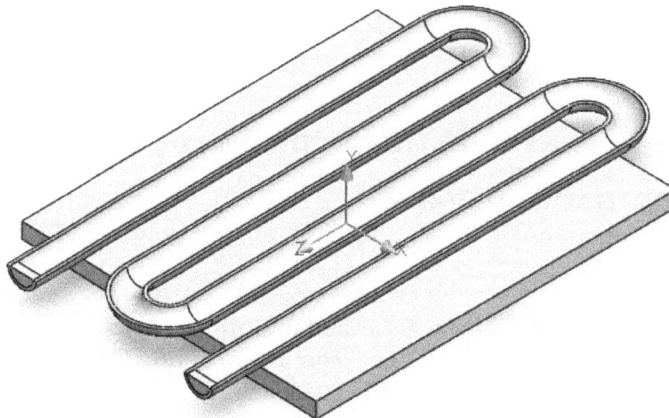

Figure 6-20 *Section view of model*

Fluid Subdomain

Next, you need to insert the fluid subdomain.

1. Right-click on the **Fluid Subdomains** in the **Flow Simulation Analysis** tree; a shortcut menu is displayed. Select the **Insert Fluid Subdomain** option from it; the **Fluid Subdomain PropertyManager** is displayed, refer to Figure 6-21.

2. Select the inside face of the pipe from the graphics area; the selected face is displayed in the **Faces to Apply the Fluid Subdomain** area in the **Selection** rollout. Also, the preview of fluid subdomain is shown in Figure 6-22.

3. Choose the **OK** button to close the dialog box.

Figure 6-21 The partial view of the Fluid Subdomain dialog box

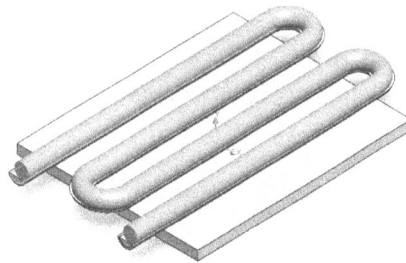

Figure 6-22 The preview of fluid subdomain

Applying Boundary Condition to Lid 1

Next, you need to invoke the **Boundary Condition** tool and apply the boundary condition to the model.

1. Right click on the **Boundary Conditions** in the **Flow Simulation Analysis** tree; a shortcut menu is displayed. Select the **Insert Boundary Condition** option from it; the **Boundary Condition PropertyManager** is displayed.

2. Select the inside face of Lid1 from the graphics area, refer to Figure 6-23; the selected face is displayed in **Faces to Apply the Boundary Condition** selection box in the **Selection** rollout.

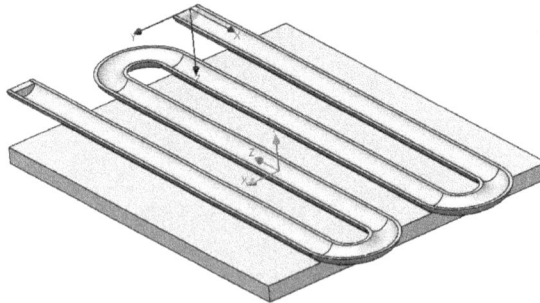

Figure 6-23 *The Lid1 face selected*

3. Select the **Inlet Mass Flow** from the **Type** rollout. Ensure that the **Flow Openings** button is chosen in the **Type** rollout while selecting the **Inlet Mass Flow**.

4. Enter **0.008** in the **Mass Flow Rate** edit box and ensure that the **Normal to Face** button is chosen in the **Flow Parameters** rollout.

5. Choose the **OK** button from the **Boundary PropertyManager** to close it.

Applying Boundary Condition to Lid 2

Next, you need to apply the inlet velocity parameter to Lid 2.

1. Right click on the **Boundary Conditions** in the **Flow Simulation Analysis** tree; a shortcut menu is displayed. Select the **Insert Boundary Condition** option from it; the **Boundary Condition PropertyManager** is displayed.

2. Select the inside face of Lid 2 from the graphics area; the selected face is displayed in the **Faces to Apply the Boundary Condition** selection box in the **Selection** rollout, refer to Figure 6-24.

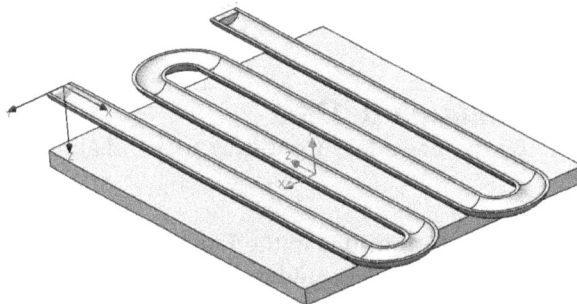

Figure 6-24 *The Lid 2 face selected*

3. Select the **Static Pressure** from the **Type** rollout. Ensure that the **Pressure Openings** button is selected in the **Type** rollout while selecting the **Static Pressure**.

4. Enter **-12 °c** in the **Temperature** edit box in the **Thermodynamic Parameters** rollout.

5. Choose the **OK** button from the **Boundary PropertyManager** to close it.

Applying the Surface Source

Next, you need to apply the surface source to the back face of plate.

1. Right-click on the **Heat Sources** in the **Flow Simulation Analysis** tree; a shortcut menu is displayed. Select the **Insert Surface Source** option from it; the **Surface Source PropertyManager** is displayed.

2. Select the back face of plate from the graphics area, refer to Figure 6-25; the selected face is displayed in the **Faces to Apply the Boundary Condition** selection box in the **Selection** rollout.

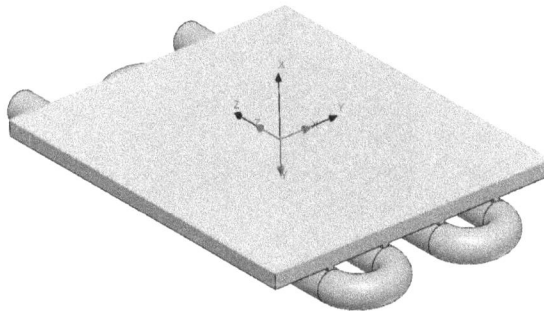

Figure 6-25 *The face selected for heat source*

3. Select the **Heat Generation Rate** button from the **Parameter** rollout.

4. Enter **1200** in the **Heat Generation Rate** edit box in the **Parameter** rollout.

5. Choose the **OK** button from the **Surface Source PropertyManager** to close it.

Saving the Model

1. Save the part document with the name *c06_tut02* at the following location: *\SOLIDWORKS_ Flow\resources\c06*.

2. Choose **File > Close** from the SOLIDWORKS menus to close the document.

Tutorial 3

In this tutorial, you will open the model (*c04_tut04*) created in Tutorial 4 of Chapter 4. You can also download this file from *www.cadcim.com* by using the following path:

*Textbooks > CAE Simulation > Dassault Systemes > SOLIDWORKS Flow Simulation > Flow Simulation
Using SOLIDWORKS 2023 > Input Files > C06_SWFS_inp*

After opening the file, you will then apply the boundary conditions to the model. The model
is shown in Figure 6-26.

(Expected time: 20 min)

The following steps are required to complete this tutorial:

a.   Open Tutorial 4 of Chapter 4.
b.   Save this tutorial in the *c06* folder with a new name.
c.   Add boundary conditions to the model.
d.   Save the file.

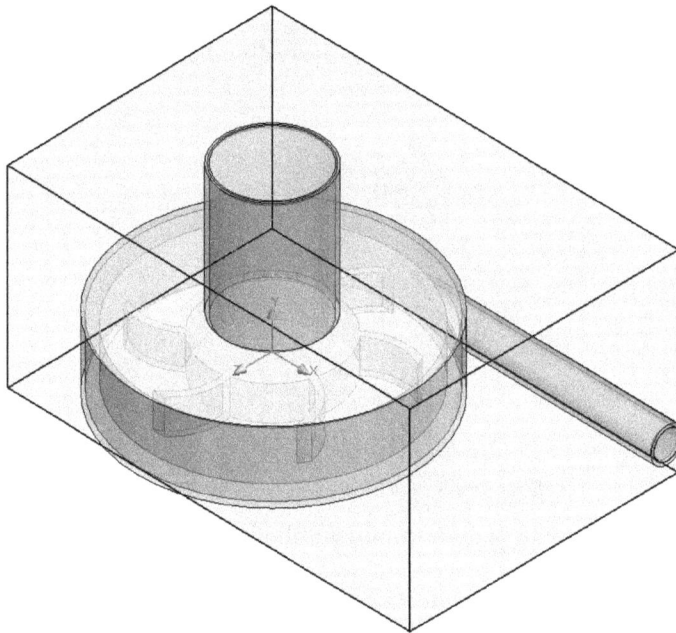

Figure 6-26 The model for Tutorial 2

Opening Tutorial 4 of Chapter 4

As the required document is saved in the *c04* folder, you need to select this folder and then
open the *c04_tut04.sldprt* document.

1.   Start SOLIDWORKS by double-clicking on its shortcut icon on the desktop of your computer.

2.   Choose the **Open** button from the Menu Bar to display the **Open** dialog box.

3.   Browse to the SOLIDWORKS folder and select the **c04** folder.

4. Select the **c04_tut04.sldprt** document and then choose the **Open** button.

Saving the Document

When you open a document from another chapter, it is recommended that you first save the opened document with a new name in the folder of the current chapter to avoid the original document from getting modified.

1. Choose the **Save As** button from the **Save** flyout in the Menu Bar; the **Save As** dialog box is displayed, as shown in Figure 6-27.

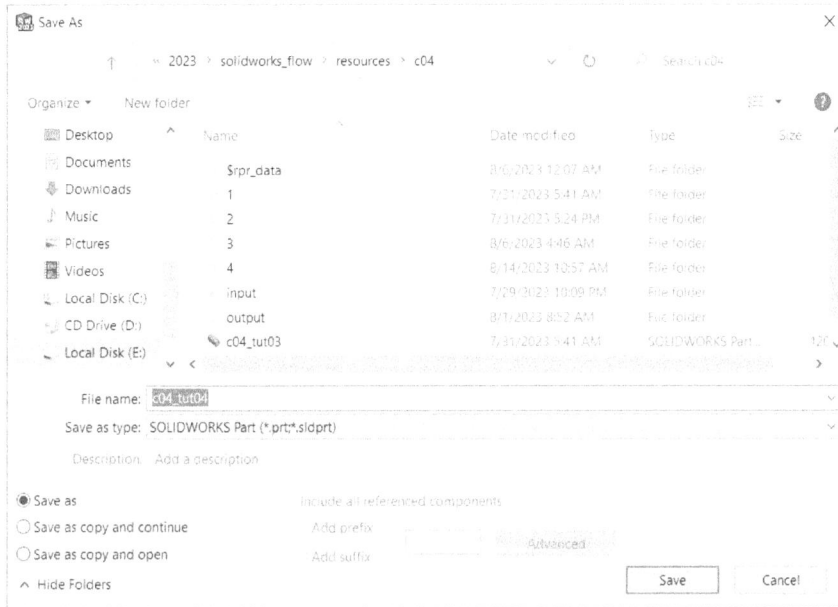

Figure 6-27 The Save As dialog box

2. Browse to the **SOLIDWORKS_Flow > Resources** folder and make the **c06** folder as the current folder by double-clicking on it.

3. Enter **c06_tut03** as the new name of the document in the **File name** edit box and then choose the **Save** button to save the document.

4. The document is saved with the new name and gets opened in the drawing area.

Editing the Computational Domain

Next, you need to edit the computational domain from the flow simulation analysis tree.

1. Select the **Computational Domain** from the **Flow Simulation Design** tree and right-click on it; a shortcut menu is displayed. Select the **Edit Definition** option from the shortcut menu; the **Computational Domain PropertyManager** is displayed, as shown in Figure 6-28.

2. Enter the following values in the **Size and Conditions** rollout, as shown in Figure 6-28.

3. Choose the **OK** button to close the PropertyManager.

    After specifying the values, the computational domain size will look like shown in Figure 6-29.

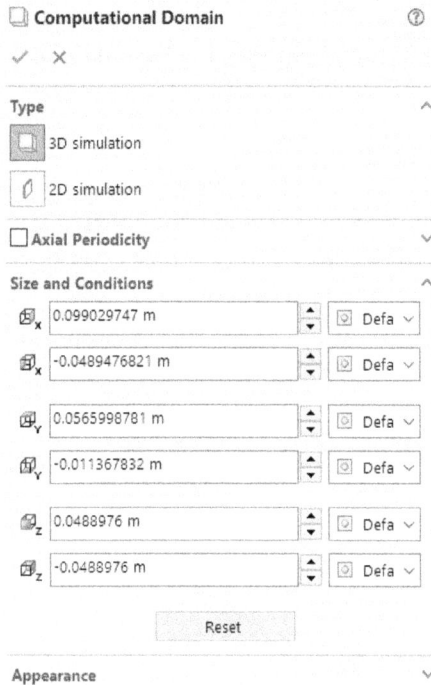

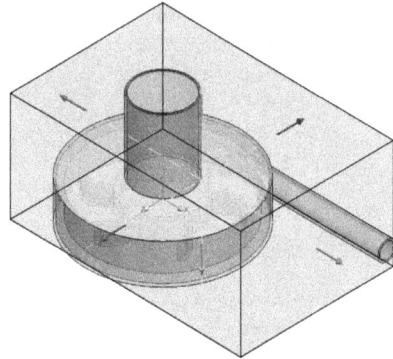

Figure 6-28 *The **Computational Domain** dialog box* ***Figure 6-29*** *The Computational domain boundary*

Hiding the Computational Domain

Next, you need to suppress the computational domain from the flow simulation analysis tree.

1. Select the **Computational Domain** from the **Flow Simulation Design** tree and right-click on it; a shortcut menu is displayed. Choose the **Hide** option from the shortcut menu, the computational domain gets hidden in the graphics area. Refer to Figure 6-30 after hiding the computational domain.

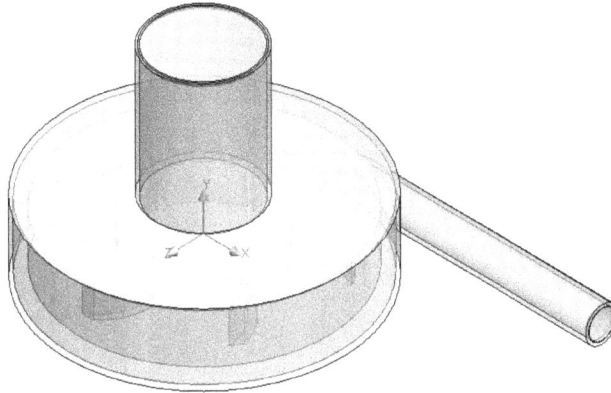

Figure 6-30 *The model without computational domain boundary*

Assign Rotating Region

Next, you need to specify the rotating region.

1. Right-click on the **Rotating Regions** in the flow simulation analysis tree; a shortcut menu is displayed. Select the **Insert Rotating Region** option from it; the **Rotating Region 1 PropertyManager** is displayed, refer to Figure 6-31.

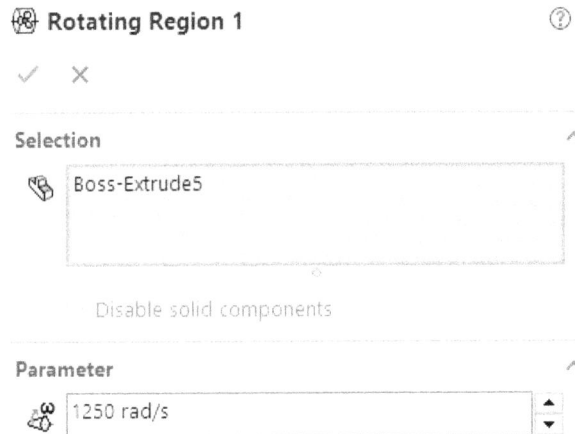

Figure 6-31 *The **Rotating Region 1 PropertyManager***

2. Select the rotating body from the graphics area which will be displayed in the Component to Apply the Rotating Region area in the **Selection** rollout, refer to Figure 6-32. Enter **1250** in the angular velocity edit box in the **Parameter** rollout.

3. Choose the **OK** button to close the dialog box.

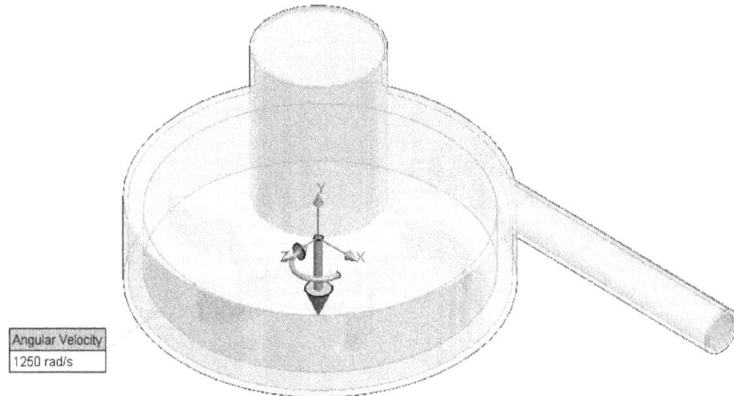

Figure 6-32 *Assign the rotational region*

Applying Boundary Condition to Lid 1

Next, you need to invoke the **Boundary Condition** tool and apply the boundary condition to the model.

1. Right-click on the **Boundary Conditions** in the **Flow Simulation Analysis** tree; a shortcut menu is displayed. Select the **Insert Boundary Condition** option from it; the **Boundary Condition PropertyManager** is displayed.

2. Select the Lid1 body from the graphics area, refer to Figure 6-33; the selected body is displayed in **Faces to Apply the Boundary Condition** selection box in the **Selection** rollout.

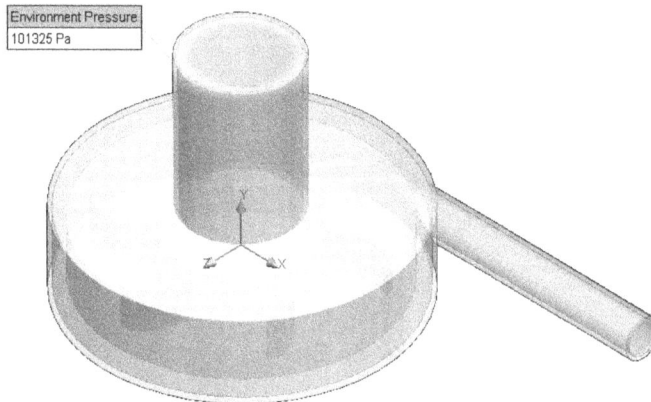

Figure 6-33 *The Lid1 selected*

3. Select the **Environment Pressure** from the **Type** rollout. Ensure that the **Pressure Openings** button is chosen in the **Type** rollout while selecting the **Environment Pressure**.

4. Choose the **OK** button from the PropertyManager to close it.

Applying Boundary Condition to Lid 2

Next, you need to apply the inlet velocity parameter to Lid 2.

1. Right click on the **Boundary Conditions** in the **Flow Simulation Analysis** tree; a shortcut menu is displayed. Select the **Insert Boundary Condition** option from it;the **Boundary Condition PropertyManager** is displayed.

2. Select the Lid2 body from the graphics area, the selected body is displayed in the **Faces to Apply the Boundary Condition** selection box in the **Selection** rollout, refer to Figure 6-34.

3. Select the **Outlet Velocity** from the **Type** rollout. Ensure that the **Flow Openings** button is selected in the **Type** rollout while selecting the **Outlet Velocity**.

4. Enter **18** in the **Velocity Normal to Face** edit box in the **Flow Parameters** rollout.

5. Choose the **OK** button from the **Boundary PropertyManager** to close it.

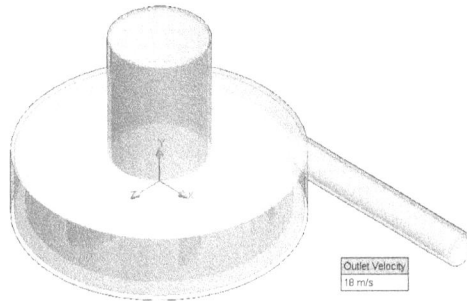

Figure 6-34 The Lid2 selected

Saving the Model

1. Save the part document with the name *c06_tut03* at the following location: *\SOLIDWORKS_ Flow\resources\c06*.

2. Choose **File > Close** from the SOLIDWORKS menus to close the document.

Self-Evaluation Test

Answer the following questions and then compare them to those given at the end of this chapter:

1. The _____ is the region where the flow and heat transfer calculations are performed.

2. The _____ button in the **Computational PropertyManager** is used to simulate 2D flow.

3. The _____ flow pattern depends on both boundary conditions and initial conditions.

4. The _____ button must be selected to define the inlet velocity.

5. The pressure of a fluid flow that is not moving but is at rest is known as _____ .

6. The _____ tool allows you to define a heat surface source on a surface.

Review Questions

Answer the following questions:

1. Which of the following PropertyManagers is displayed when you choose the **Computational Domain** tool from the **Flow Simulation CommandManager**?

    (a) **Computation**   (b) **Compute Domain**
    (c) **Computational Domain** (d) None of these

2. Which of the following parameters is not available in the flow opening type boundary condition?

    (a) **Inlet Mass Flow**   (b) **Inlet Velocity**
    (c) **Static Pressure**   (d) All of these

3. The _____ is a flow of energy per unit of area per unit of time.

4. The _____ button is used when you need to specify the swirling of the flow about an axis of the reference coordinate system.

5. The heat transfer rate between two surfaces equals ratio of the temperature difference and the total thermal resistance between them. (T/F)

Answers to Self-Evaluation Test

1. computational domain, **2. 2D simulation**, **3.** time-dependent, **4. Flow Openings**, **5.** Total Pressure, **6. Surface Source**

Chapter 7

Creating Goals

Learning Objectives

After completing this chapter, you will be able to:

- *Create global goals*
- *Create point goals*
- *Create surface goals*
- *Create volume goals*
- *Create equation goals*

GOALS

Goals are physical parameters whose convergence may represent the attainment of a steady-state solution from an engineering standpoint. It is important to note that Goals Convergence is one of the requirements for completing the calculation. Following are the goal types which you can specify:

a) Global Goal
b) Point Goal
c) Surface Goal
d) Volume Goal
e) Equation Goal

GLOBAL GOALS

CommandManager:	Flow Simulation > Goals > Global Goals
SOLIDWORKS Menus:	Tools > Flow Simulation > Insert > Global Goals
Toolbar:	Flow Simulation Goals Group > Global Goals

Global Goal is a physical parameter that is calculated across the entire computational domain. This tool allows you to specify global goals for a project. Choose the **Global Goals** tool from the **Goals** drop-down in the **Flow Simulation CommandManager**; the **Global Goals PropertyManager** is displayed, as shown in Figure 7-1. There are two rollouts in this property manager, **Parameters** and **Name Templates**, which are discussed next.

Parameters Rollout

You can select required parameters from the **Parameters** rollout. You can choose the parameter value, such as minimum (Min), average (Av), maximum (Max), or bulk average (Bulk av.) for the parameters (static pressure, total pressure, and many more).

$$\text{Average} = \sum_i \frac{A_i dV_i}{dV_i}$$

$$\text{Bulk Average} = \sum_i \frac{A_i \rho_i dV_i}{\rho_i dV_i}$$

Ai = averaged parameter (example: velocity)

dVi = volume of ith cell

 i = density in ith cell

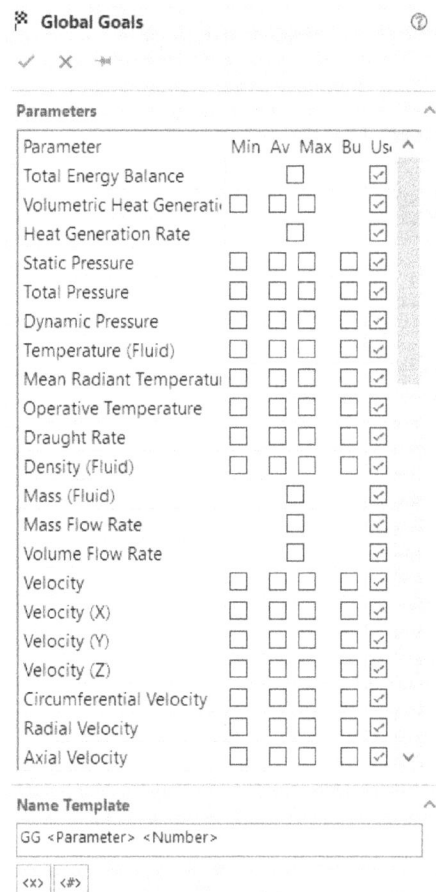

Figure 7-1 The Global Goals PropertyManager

You can select the **Use for Convergence Control(Use for Conv.)** check box if you want to keep track of the goal convergence as a requirement for completing the computation.

Name Templates

In this rollout, you can enter the name of a template in the **Name Template** edit box. The default template name is GG <Parameter> <Number>, where as GG is Global goals, <Parameter> is the goal parameter, and <Number> is a sequential number of the goal.

Choose the **OK** button after specifying the goal to close the **Global Goals PropertyManager**.

POINT GOALS

CommandManager:	Flow Simulation > Goals > Point Goals
SOLIDWORKS menus:	Tools > Flow Simulation > Insert > Point Goals
Toolbar:	Flow Simulation Goals Group > Point Goals

The point goal is the calculation of physical properties at a specified location. The **Point Goals** tool allows you to specify point goals for the project. To do so, choose the **Point Goals** tool from the **Goals** drop-down in the **Flow Simulation CommandManager**; the **Point Goals PropertyManager** is displayed, as shown in Figure 7-2. There are four rollouts in this property manager, **Points**, **Options**, **Parameters** and **Name Templates**, which are discussed next.

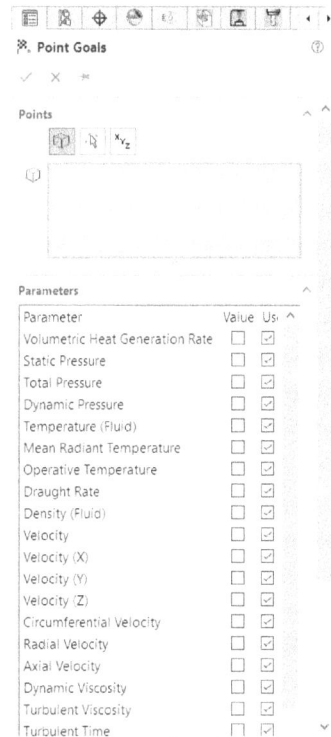

Points Rollout

The options under this are discussed next.

Reference

By using this button, you can select a point from the existing body. The selected point will be displayed in the **Points**, **Vertices**, **Planar Faces**, and **Edges and Components** lists.

Pick from Screen

By using this button, you can select points on a plane or face. The selected face will be displayed in the **Plane or Planar Face** list.

Coordinates

Figure 7-2 The partial Point Goals PropertyManager

By using this button, you can specify set of points with point coordinates. You can specify the coordinate value in the **X Coordinate**, **Y Coordinate** and **Z Coordinate** edit boxes. To add the points, choose the **Add point** button; the point will be added in the **Coordinates in the Global Coordinate System** list.

The other rollouts, **Parameters** and **Name Template**, are already discussed previously.

Choose the **OK** button after specifying the goal to close the **Point Goals PropertyManager**.

SURFACE GOALS

CommandManager:	Flow Simulation > Goals > Surface Goals
SOLIDWORKS menus:	Tools > Flow Simulation > Insert > Surface Goals
Toolbar:	Flow Simulation Goals Group > Surface Goals

Surface goals are physical parameters calculated on the surfaces that have been selected. This tool allows you to specify surface goals for the project. To do so, choose the **Surface Goals** tool from the **Goals** drop-down in the **Flow Simulation CommandManager**; the **Surface Goals PropertyManager** is displayed, as shown in Figure 7-3. There are three rollouts in this property manager namely **Selection**, **Parameters**, and **Name Templates**, which are discussed next.

Selection Rollout
As you select a surface from the existing model, the selected surface will appear in the **Faces to Apply the Surface Goal** selection box. If you select multiple surfaces then the **Create goal for each surface** check box will be available which helps you to assign different goals for different surfaces. The other rollouts **Parameters** and **Name Template** are already discussed. Choose the **OK** button after specifying the goal to close the **Surface Goals PropertyManager**.

VOLUME GOALS

CommandManager:	Flow Simulation > Goals > Volume Goals
SOLIDWORKS menus:	Tools > Flow Simulation > Insert > Volume Goals
Toolbar:	Flow Simulation Goals Group > Volume Goals

A volume goal is a physical parameter calculated within specified volumes inside the Computational Domain. This tool allows you to specify volume goals for the project. To do so, choose the **Volume Goals** tool from the **Goals** drop-down in the **Flow Simulation CommandManager**; the **Volume Goals PropertyManager** is displayed, as shown in Figure 7-4. There are three rollouts in this property manager, **Selection**, **Parameters**, and **Name Templates**, which are discussed next.

Selection Rollout
As you select a body from the graphics area, the selected body will appear in the **Components to Apply the Volume Goal** selection box. If you select multiple bodies then the **Create goal for each component** check box will be available which helps you to assign different goals for different bodies. The other rollouts, **Parameters** and **Name Template**, are already discussed previously. Choose the **OK** button after specifying the goal to close the **Volume Goals PropertyManager**.

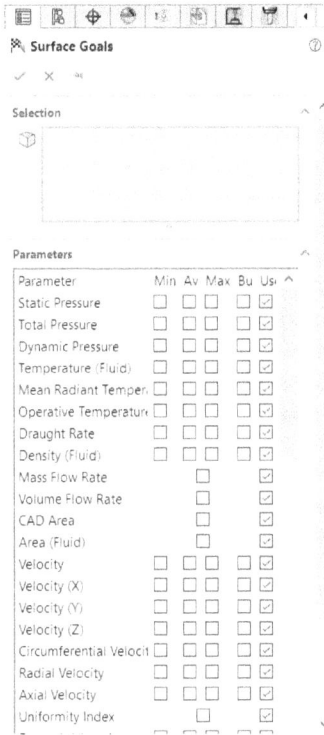

Figure 7-3 *The partial view of* **Surface** *Goals PropertyManager*

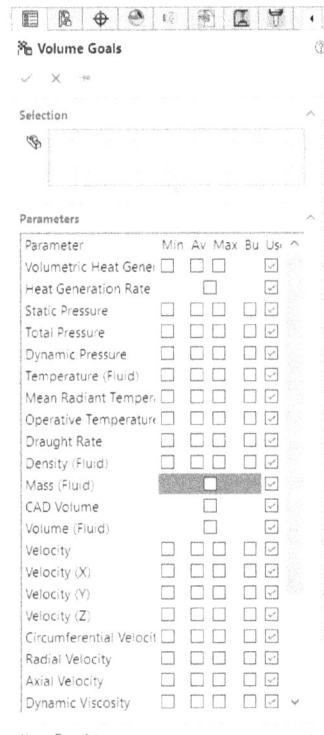

Figure 7-4 *The partial view of* **Volume** *Goals PropertyManager*

EQUATION GOALS

CommandManager:	Flow Simulation > Goals > Equation Goal
SOLIDWORKS menus:	Tools > Flow Simulation > Insert > Equation Goal
Toolbar:	Flow Simulation Goals Group > Equation Goal

An equation goal is a physical parameter which is defined using equations. This tool allows you to specify equation goals for a project. To do so, choose the **Equation Goal** tool from the **Goals** drop-down in the **Flow Simulation CommandManager**; the **Equation Goal** tab is displayed at the lower left corner of the drawing area, as shown in Figure 7-5. Next, complete the equation definition in the **Expression** area using the calculator style buttons. Choose the correct units for the equation from the **Dimensionality** drop-down list. Next, choose the **OK** button after specifying the goal to close the **Equation Goal** tab.

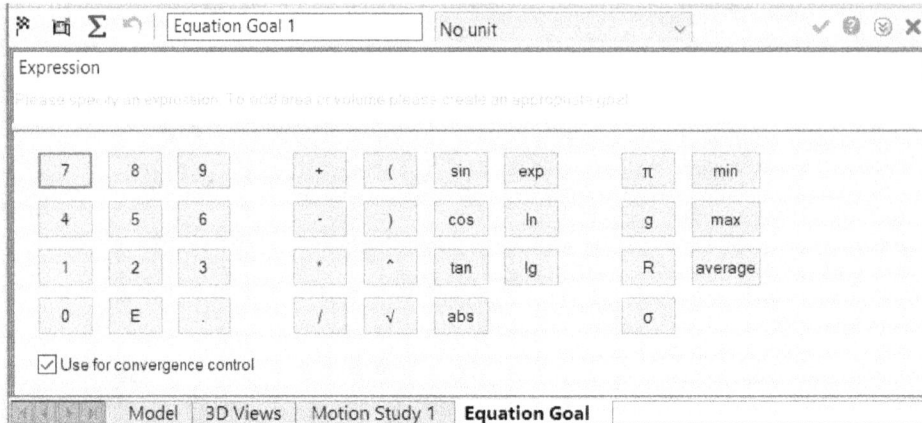

*Figure 7-5 The **Equation Goal** tab*

TUTORIALS

Tutorial 1

In this tutorial, you will open the model (*c06_tut01*) created in Tutorial 1 of Chapter 6. You can also download this file from *www.cadcim.com* by using the following path:

Textbooks > CAE Simulation > Dassault Systemes > SOLIDWORKS Flow Simulation > Flow Simulation Using SOLIDWORKS 2023 > Input Files > C07_SWFS_inp

You will then apply goal to the model. The model is shown in Figure 7-6.

(Expected time: 20 min)

The following steps are required to complete this tutorial:

a. Open Tutorial 1 of Chapter 6.
b. Save this tutorial in the *c07* folder with a new name.
c. Add goals to the model.
d. Save the file.

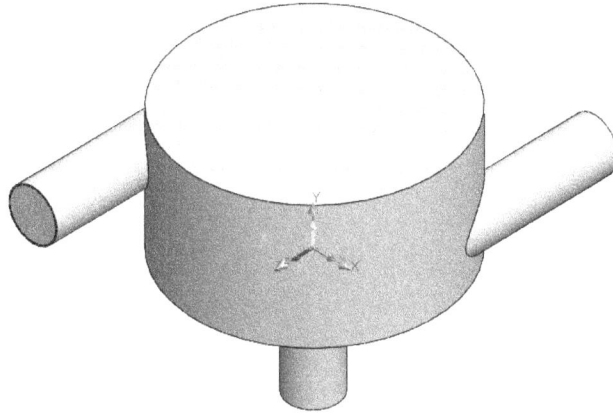

Figure 7-6 *Model for Tutorial 1*

Opening Tutorial 1 of Chapter 6

As the required tutorial is saved in the *c06* folder, you need to select this folder and then open the *c06_tut01.sldprt* document.

1. Start SOLIDWORKS by double-clicking on its shortcut icon on the desktop of your computer.

2. Choose the **Open** button from the Menu Bar to display the **Open** dialog box, refer to Figure 7-7.

3. Browse to the SOLIDWORKS folder and select the **c06** folder.

4. Select the **c06_tut01.sldprt** document and then choose the **Open** button.

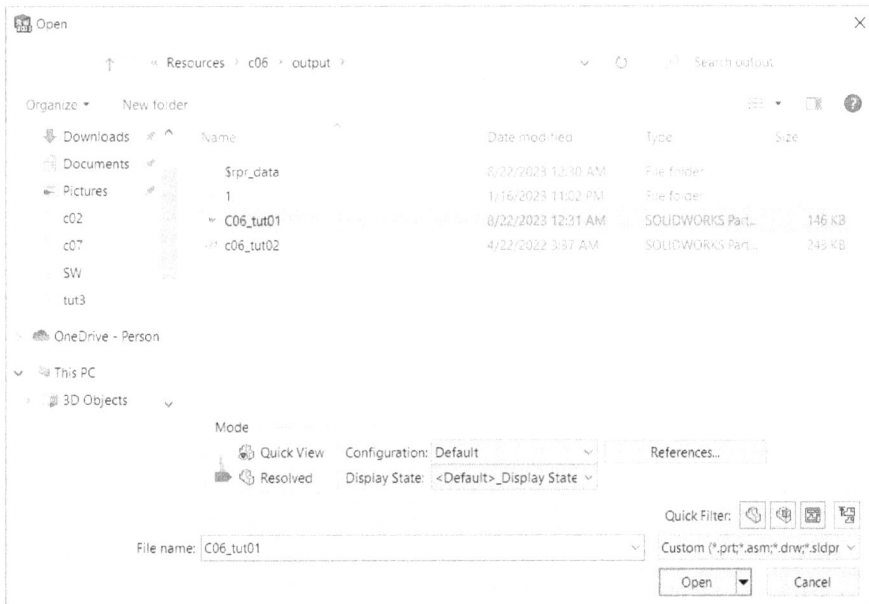

Figure 7-7 *The **Open** dialog box*

As the model was saved in the Flow Simulation environment in Chapter 6, it opens in the Flow Simulation environment, refer to Figure 7-8.

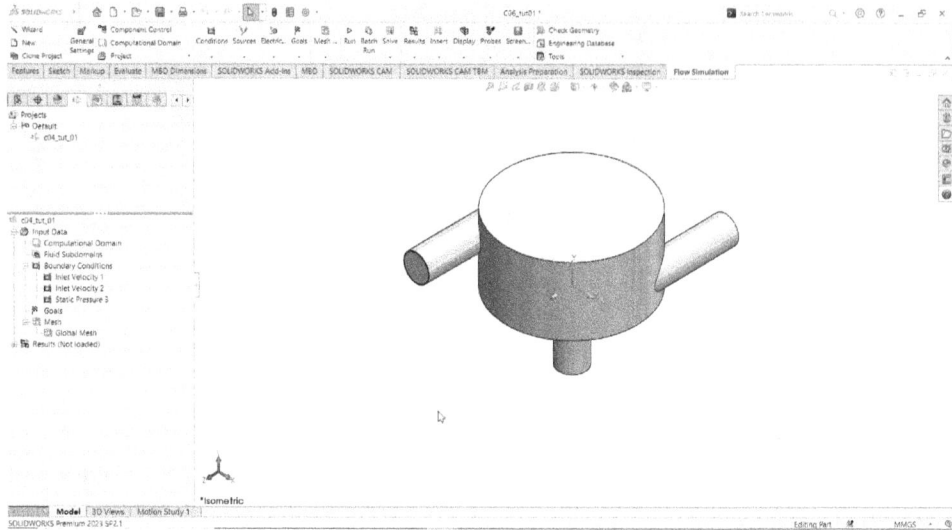

Figure 7-8   *The model opened in the Flow Simulation environment*

Saving the Document in the c07 Folder

When you open a document from another chapter, it is recommended to first save the opened document with a new name in the folder of the current chapter to avoid the original document from getting modified.

1. Choose the **Save As** button from the **Save** flyout in the Menu Bar; the **Save As** dialog box is displayed, refer to Figure 7-9.

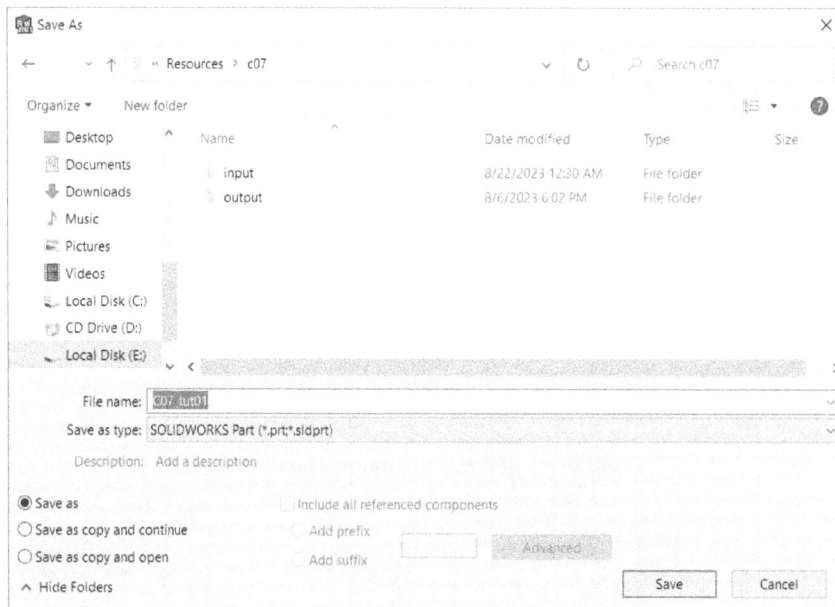

Figure 7-9   *The **Save As** dialog box*

2. Browse to the **SOLIDWORKS_Flow > Resources** folder and then create a new folder with the name **c07** by using the **Create New Folder** button. Make the **c07** folder as the current folder by double-clicking on it.

3. Enter **c07_tut01** as the new name of the document in the **File name** edit box and then choose the **Save** button to save the document.

    The document is saved with the new name and gets opened in the drawing area.

Applying Goal to Lid 3

Next, you need to invoke the **Surface Goals** tool and apply the boundary condition to the model.

1. Choose the **Surface Goals** tool from the **Goals** drop-down in the **Flow Simulation CommandManager**; the **Surface Goals PropertyManager** is displayed, as shown in Figure 7-10.

2. Select the inner face of Lid3 from the graphics area, refer to Figure 7-11. The selected face is displayed in the **Faces to Apply the Surface Goal** selection box in the **Selection** rollout.

3. Select the **Average** check box to the right of the **Velocity** parameter.

4. Select the **Average** check box to the right of the **Temperature (Fluid)** parameter.

5. Choose the **OK** button from the **Surface Goals PropertyManager** to close it.

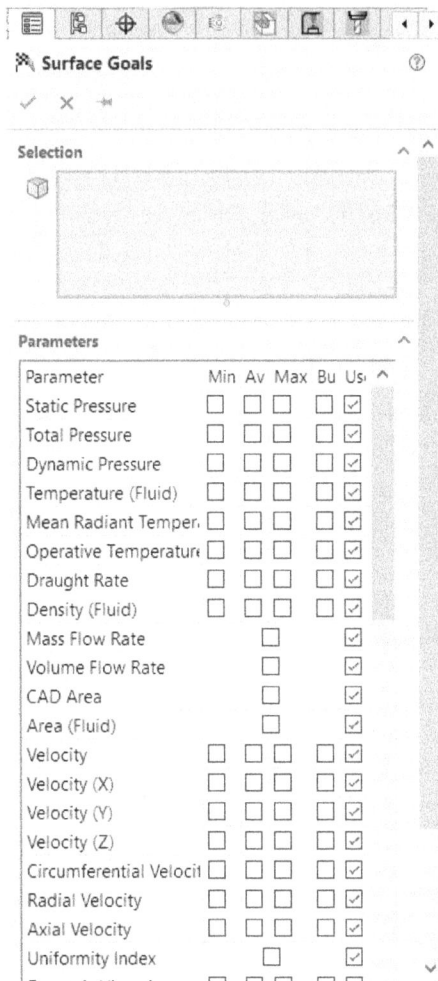

Figure 7-10 *The partial view of the*
Surface Goals PropertyManager

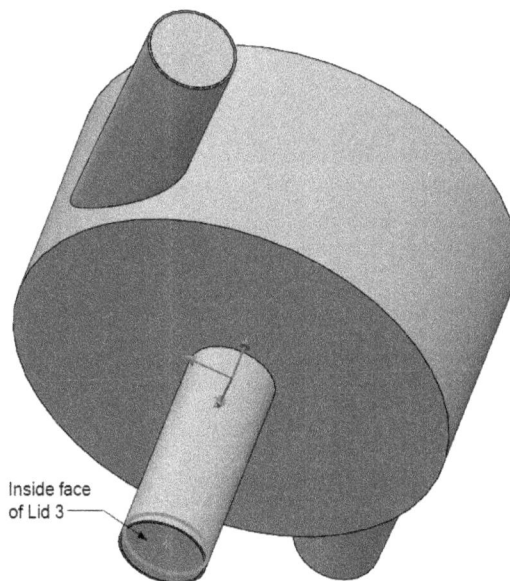

Figure 7-11 *The inside face of Lid 3*

Saving the Model

1. Save the part document with the name **c07_tut01** at the following location: *\SOLIDWORKS_ Flow\resources\c07*.

2. Choose **File > Close** from the SOLIDWORKS menus to close the document.

Tutorial 2

In this tutorial, you will open the model (*c06_tut02*) created in Tutorial 2 of Chapter 6. You can also download this file from *www.cadcim.com* by using the following path:

Textbooks > CAE Simulation > Dassault Systemes > SOLIDWORKS Flow Simulation > Flow Simulation Using SOLIDWORKS 2023 > Input Files > C07_SWFS_inp

You will then apply goals to the model. The model is shown in Figure 7-12.

(Expected time: 20 min)

The following steps are required to complete this tutorial:

a. Open *c06_tut02*.
b. Save this tutorial in the *c07* folder with a new name.
c. Add goals to the model.
d. Save the file.

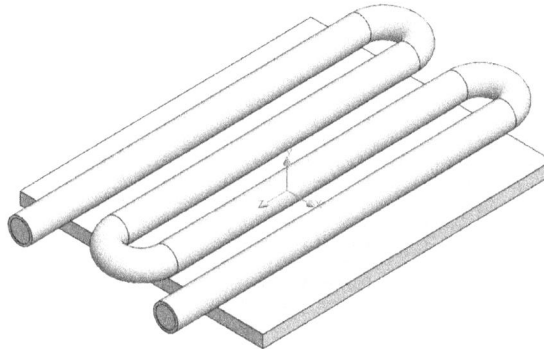

Figure 7-12 Model for Tutorial 2

Opening Tutorial 2 of Chapter 6

As the required document is saved in the c06 folder, you need to select this folder and then open the *c06_tut02.sldprt* document.

1. Start SOLIDWORKS by double-clicking on its shortcut icon on the desktop of your computer.

2. Choose the **Open** button from the Menu Bar to display the **Open** dialog box, refer to Figure 7-13.

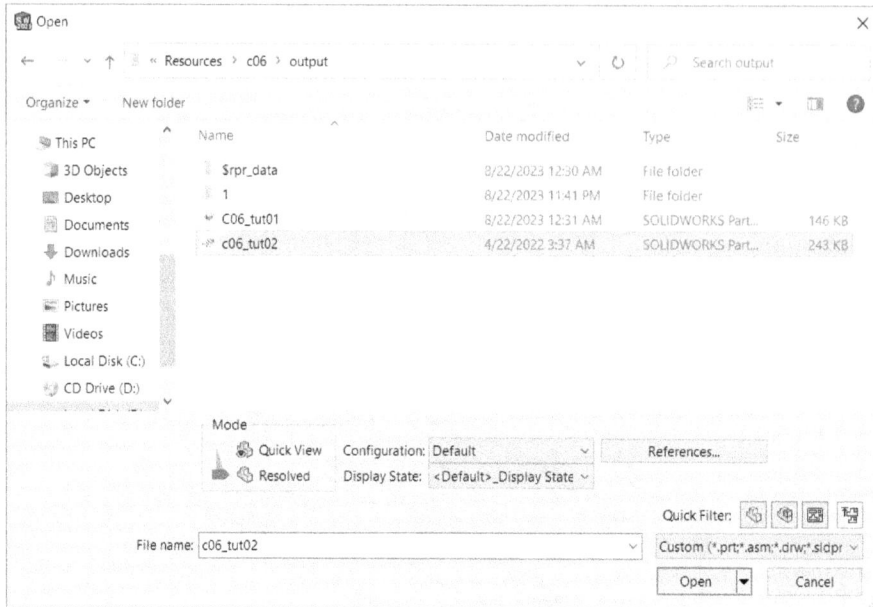

*Figure 7-13 The **Open** dialog box*

3. Browse to the SOLIDWORKS folder and select the **c06** folder.

4. Select the **c06_tut02.sldprt** document and then choose the **Open** button. The model is opened in the Flow Simulation environment, refer to Figure 7-14.

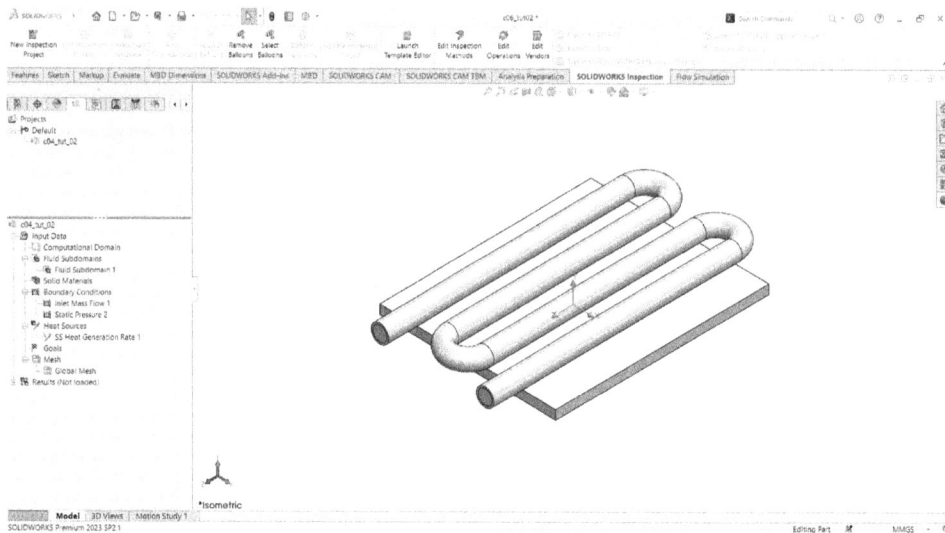

Figure 7-14 The model opened in the Flow Simulation environment

Saving the Document

When you open a document from another chapter, it is recommended that you first save the opened document with a new name in the folder of the current chapter to avoid the original document from getting modified.

1. Choose the **Save As** button from the **Save** flyout in the Menu Bar; the **Save As** dialog box is displayed, as shown in Figure 7-15.

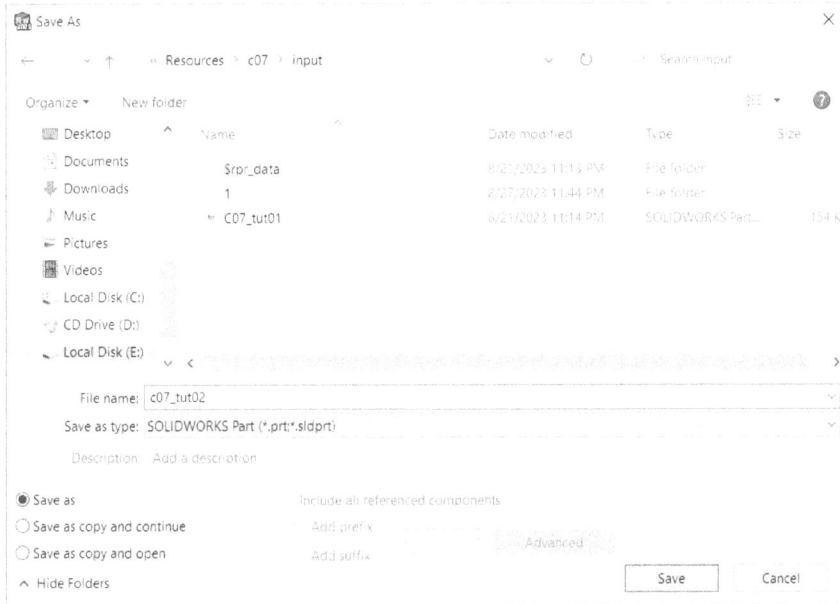

Figure 7-15   The Save As dialog box

2. Browse to the **SOLIDWORKS_Flow > Resources** folder and make the **c07** folder as the current folder by double-clicking on it.

3. Enter **c07_tut02** as the new name of the document in the **File name** edit box and then choose the **Save** button to save the document.

4. The document is saved with the new name and gets opened in the drawing area.

Applying Goal to Lid 2

Next, you need to invoke the **Surface Goals** tool and apply the boundary condition to the model.

1. Choose the **Surface Goals** tool from the **Goals** drop-down in the **Flow Simulation CommandManager**; the **Surface Goals PropertyManager** is displayed, as shown in Figure 7-16.

2. Select the inside face of Lid 2 from the graphics area, refer to Figure 7-17; the selected face is displayed in the **Faces to Apply the Surface Goal** selection box in the **Selection** rollout.

3. Select the **Maximum** check box to the right of the **Temperature(Fluid)** parameter.

4. Choose the **OK** button from the **Surface Goals PropertyManager** to close it.

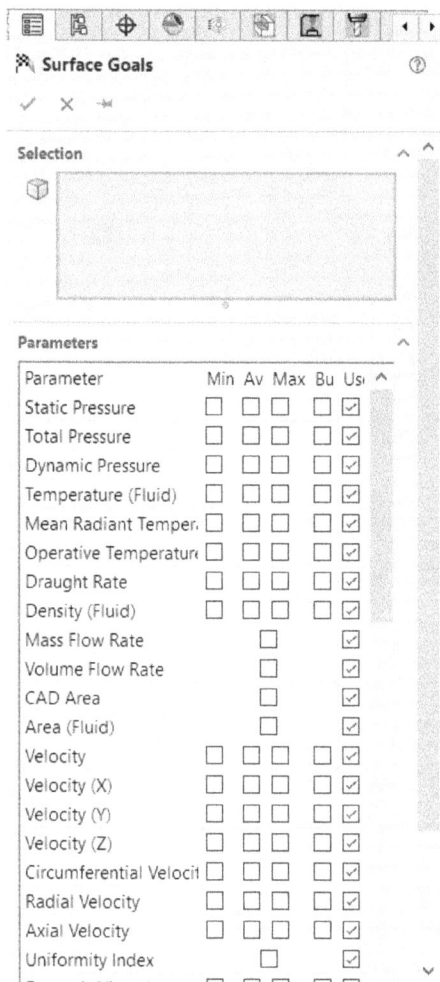

Figure 7-16 The partial view of *Surface Goals PropertyManager*

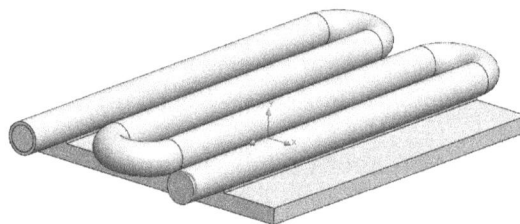

Figure 7-17 The inside face of Lid 2

Applying a Goal to Source Plate

Next, you need to invoke the Surface Goals tool and apply the boundary condition to the model.

1. Choose the **Surface Goals** tool from the **Goals** drop-down in the **Flow Simulation CommandManager**; the **Surface Goals PropertyManager** is displayed.

2. Select the bottom face of the plate from the graphics area, refer to Figure 7-18; the selected face is displayed in the **Faces to Apply the Surface Goal** selection box in the **Selection** rollout.

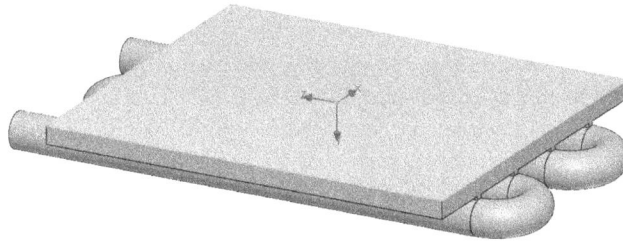

Figure 7-18 *The bottom face of the plate*

3.   Select the **Maximum** check box available on the right of the **Total Temperature** parameter.

4.   Choose the **OK** button from the **Surface Goals PropertyManager** to close it.

Applying Global Goals

Next, you need to invoke the **Global Goals** tool and apply the boundary condition to the model.

1.   Choose the **Global Goals** tool from the **Goals** drop-down in the **Flow Simulation CommandManager**; the **Global Goals PropertyManager** is displayed, as shown in Figure 7-19.

2.   Select the **Maximum** check box available on the right of the **Temperature(Fluid)** parameter.

3.   Select the **Maximum** check box available on the right of the **Total Temperature** parameter.

4.   Choose the **OK** button from the **Global Goals PropertyManager** to close it.

Saving the Model

1.   Save the part document with the name *c07_tut02* at the following location: *SOLIDWORKS_Flow\resources\c07*.

2.   Choose the **File > Close** from the SOLIDWORKS menus to close the document.

Figure 7-19 *The Global Goals PropertyManager*

Tutorial 3

In this tutorial, you will open the model (*c04_tut03*) created in Tutorial 3 of Chapter 4. You can also download this file from *www.cadcim.com* by using the following path:

Textbooks > CAE Simulation > Dassault Systemes > SOLIDWORKS Flow Simulation > Flow Simulation Using SOLIDWORKS 2023 > Input Files > C07_SWFS_inp

You will then apply goals to the model. The model is shown in Figure 7-20.

(Expected time: 20 min)

The following steps are required to complete this tutorial:

a. Open *c04_tut03*.
b. Save this tutorial in the *c07* folder with a new name.
c. Add goals to the model.
d. Save the file.

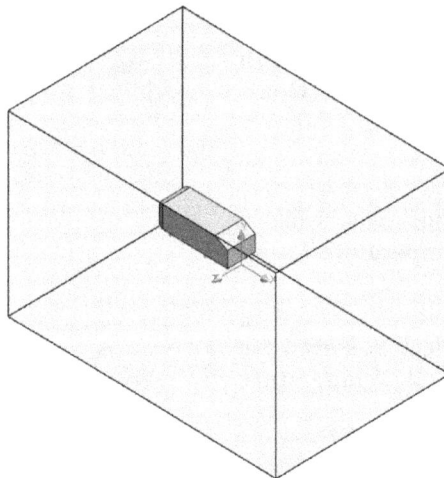

Figure 7-20 Model for Tutorial 3

Opening Tutorial 2 of Chapter 4

As the required document is saved in the *c04* folder, you need to select this folder and then open the *c04_tut03.sldprt* document.

1. Start SOLIDWORKS by double-clicking on its shortcut icon on the desktop of your computer.

2. Choose the **Open** button from the Menu Bar to display the **Open** dialog box, refer to Figure 7-21.

3. Browse to the SOLIDWORKS folder and select the **c04** folder.

4. Select the **c04_tut03.sldprt** document and then choose the **Open** button. The model is opened in the Flow Simulation environment, refer to Figure 7-22.

*Figure 7-21   The **Open** dialog box*

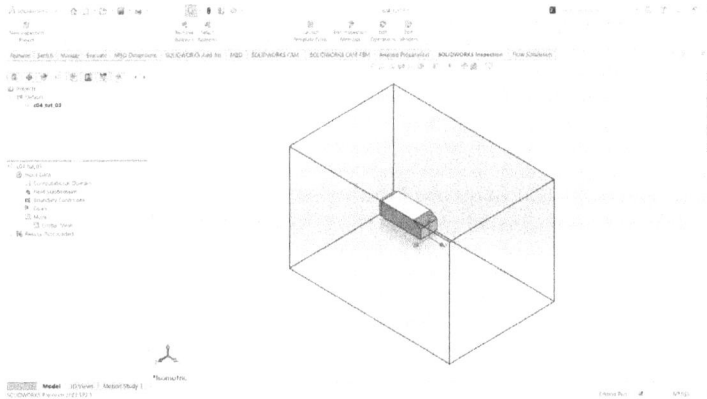

Figure 7-22   The model opened in the Flow Simulation environment

Saving the Document

When you open a document from another chapter, it is recommended that you first save the opened document with a new name in the folder of the current chapter to avoid the original document from getting modified.

1. Choose the **Save As** button from the **Save** flyout in the Menu Bar; the **Save As** dialog box is displayed, as shown in Figure 7-23.

2. Browse to the **SOLIDWORKS_Flow > Resources** folder and make the **c07** folder as the current folder by double-clicking on it.

3. Enter **c07_tut03** as the new name of the document in the **File name** edit box and then choose the **Save** button to save the document.

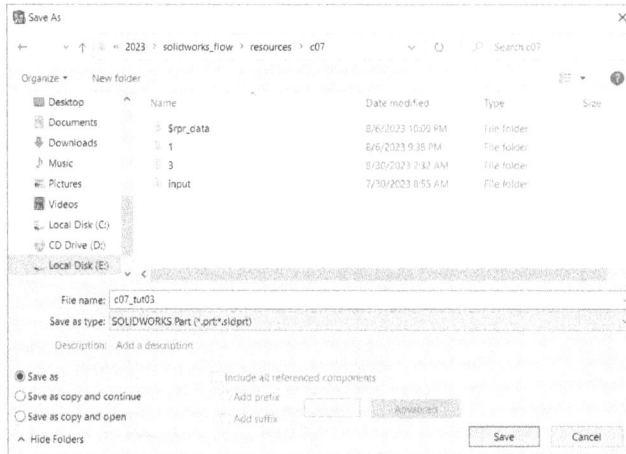

Figure 7-23 The *Save As* dialog box

4. The document is saved with the new name and gets opened in the drawing area.

Applying Global Goals

Next, you need to invoke the **Global Goals** tool and apply the boundary condition to the model.

1. Choose the **Global Goals** tool from the **Goals** drop-down in the **Flow Simulation CommandManager**; the **Global Goals PropertyManager** is displayed, as shown in Figure 7-24.

2. Select the **Maximum** check box available on the right of the **Force(X)** parameter.

3. Choose the **OK** button from the **Global Goals PropertyManager** to close it.

Applying Equation Goal

Next, you need to invoke the **Equation Goal** tool and apply the boundary condition to the model.

1. Choose the **Equation Goal** tool from the **Goals** drop-down in the **Flow Simulation CommandManager**; the **Equation Goal** tab is displayed, as shown in Figure 7-25.

2. Enter the equation in the **Expression** area, refer to Figure 7-25.

3. Choose the **OK** button from the **Equation Goal** tab to close it.

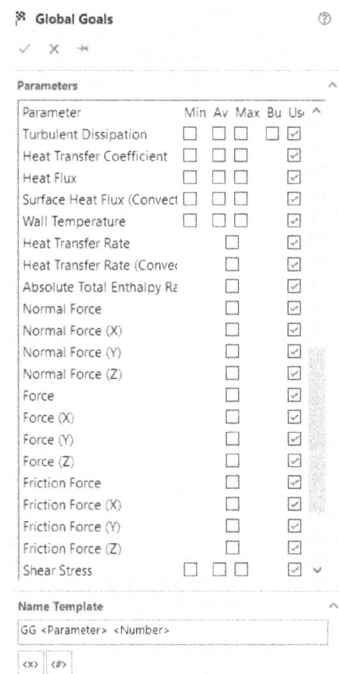

Figure 7-24 The *Global Goals* PropertyManager

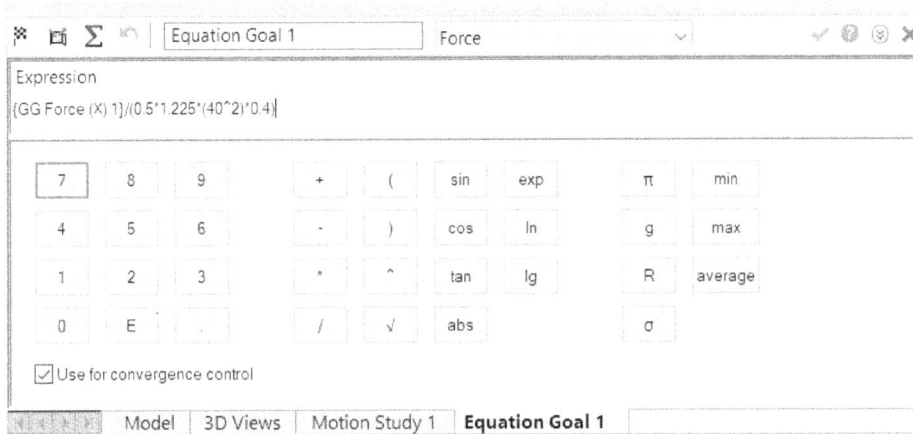

Figure 7-25 *The* **Equation Goal** *tab*

Saving the Model

1. Save the part document with the name **c07_tut03** at the following location: *SOLIDWORKS_ Flow\resources\c07*.

2. Choose **File > Close** from the SOLIDWORKS menus to close the document.

Tutorial 4

In this tutorial, you will open the model (*c06_tut03*) created in Tutorial 3 of Chapter 6. You can also download this file from *www.cadcim.com* by using the following path:

Textbooks > CAE Simulation > Dassault Systemes > SOLIDWORKS Flow Simulation > Flow Simulation Using SOLIDWORKS 2023 > Input Files > C07_SWFS_inp

You will then apply goals to the model. The model is shown in Figure 7-26.

(Expected time: 20 min)

The following steps are required to complete this tutorial:

a. Open Tutorial 3 of Chapter 6.
b. Save this tutorial in the *c07* folder with a new name.
c. Add goals to the model.
d. Save the file.

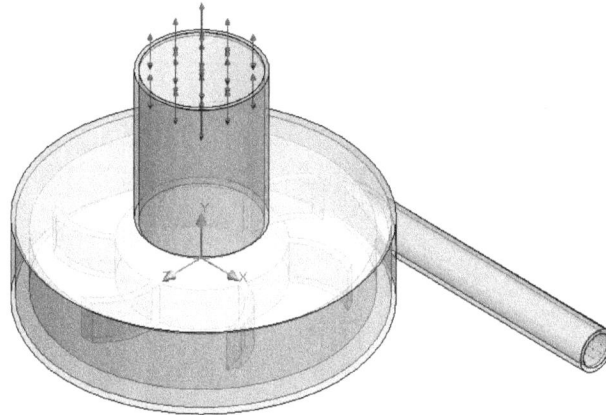

Figure 7-26 Model for Tutorial 4

Opening Tutorial 3 of Chapter 6

As the required document is saved in the *c06* folder, you need to select this folder and then open the *c06_tut03.sldprt* document.

1. Start SOLIDWORKS by double-clicking on its shortcut icon on the desktop of your computer.

2. Choose the **Open** button from the Menu Bar to display the **Open** dialog box, refer to Figure 7-27.

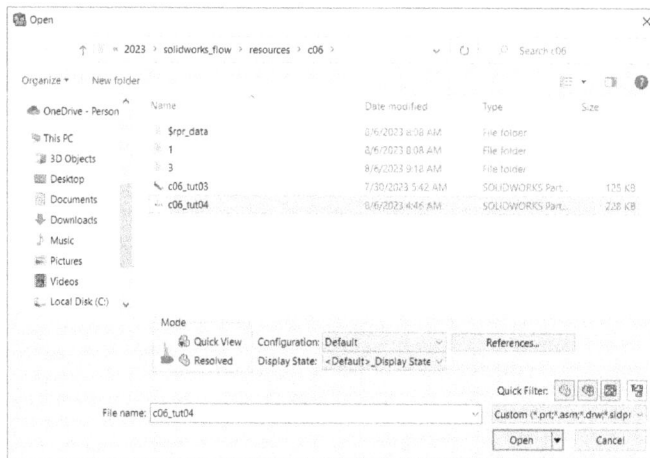

*Figure 7-27 The **Open** dialog box*

3. Browse to the SOLIDWORKS folder and select the *c06* folder.

4. Select the **c06_tut03.sldprt** document and then choose the **Open** button. The model opened in the Flow Simulation environment, refer to Figure 7-28.

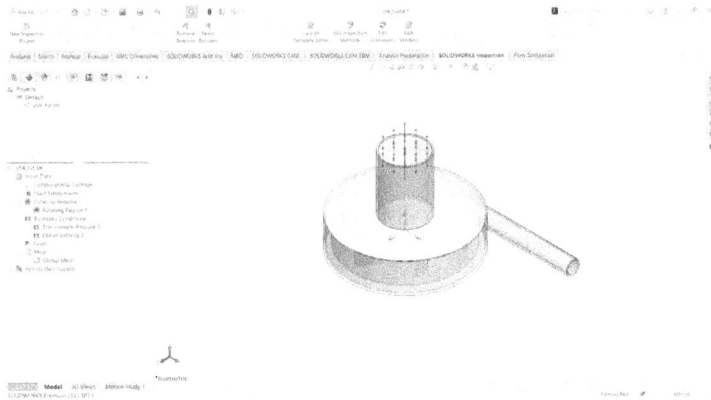

Figure 7-28 The model opened in the Flow Simulation environment

Saving the Document

When you open a document from another chapter, it is recommended that you first save the opened document with a new name in the folder of the current chapter to avoid the original document from getting modified.

1. Choose the **Save As** button from the **Save** flyout in the Menu Bar; the **Save As** dialog box is displayed, as shown in Figure 7-29.

2. Browse to the **SOLIDWORKS_Flow > Resources** folder and make the **c07** folder as the current folder by double-clicking on it.

3. Enter **c07_tut04** as the new name of the document in the **File name** edit box and then choose the **Save** button to save the document.

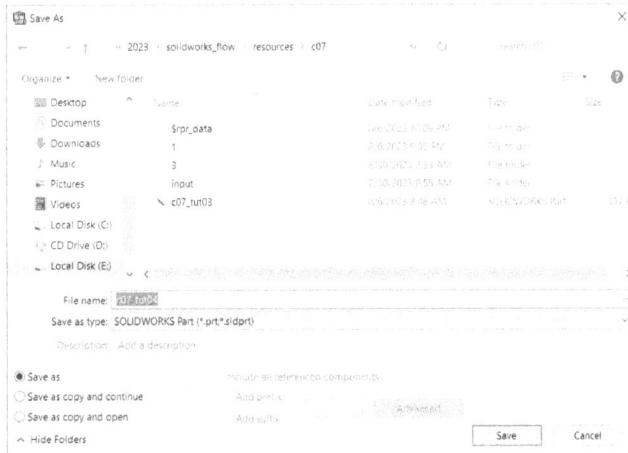

Figure 7-29 The Save As dialog box

4. The document is saved with the new name and gets opened in the drawing area.

Applying a Goal to Lid 2

Next, you need to invoke the **Surface Goals** tool and apply the goal to the Lid 2.

1. Choose the **Surface Goals** tool from the **Goals** drop-down in the **Flow Simulation CommandManager**; the **Surface Goals PropertyManager** is displayed.

2. Select the Lid 2 from the graphics area, refer to Figure 7-30; the Lid 2 is displayed in the **Faces to Apply the Surface Goal** selection box in the **Selection** rollout.

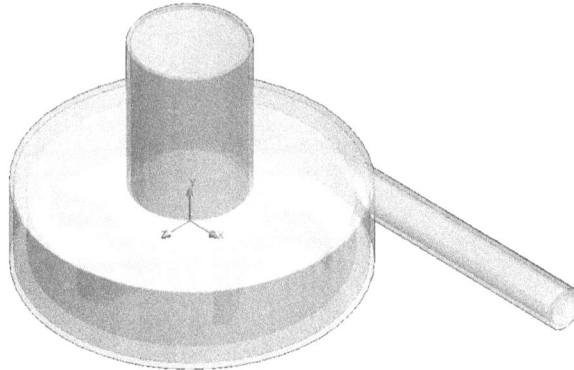

Figure 7-30 The Lid 2 face

3. Select the **Average** check box available on the right of the **Total Pressure** parameter.

4. Choose the **OK** button from the **Surface Goals PropertyManager** to close it.

Saving the Model

1. Save the part document with the name *c07_tut04* at the following location: *SOLIDWORKS_ Flow\resources\c07*.

2. Choose **File > Close** from the SOLIDWORKS menus to close the document.

Self-Evaluation Test

Answer the following questions and then compare them to those given at the end of this chapter:

1. Specifying Goals allows you to shorten the _____ time.

2. The _____ goal is a physical parameter calculated within the entire computation domain.

3. The _____ box is used to specify a template for the names of goals.

4.   A _____ is a physical parameter value calculated at a selected point.

5.   There are _____ ways to specify points for point goals.

Review Questions

Answer the following questions:

1.   Which of the following PropertyManagers is displayed when you choose the **Global Goals** tool from the **Flow Simulation CommandManager**?

    (a) **Global**   (b) **Goals**
    (c) **Global Goals**   (d) None of these

2.   Which of the following PropertyManagers is displayed when you choose the Volume Goals tool from the Flow Simulation CommandManager?

    (a) **Volume**   (b) **Vol Goals**
    (c) **Volume Goals**   (d) None of these

3.   The _____ button is used to select a point from the existing body.

4.   The _____ button is used to specify a set of points with point coordinates.

5.   Surface Goal is a physical parameter calculated on the selected surfaces. (T/F)

EXERCISES

Exercise 1

In this exercise, you will open the model (*c04_exr01*) created in Exercise 1 of Chapter 4. You can also download this file from *www.cadcim.com* by using the following path:

Textbooks > CAE Simulation > Dassault Systemes > SOLIDWORKS Flow Simulation > Flow Simulation Using SOLIDWORKS 2023 > Input Files > C07_SWFS_inp

You will then apply goals to the model. The model is shown in Figure 7-31.

(Expected time: 20 min)

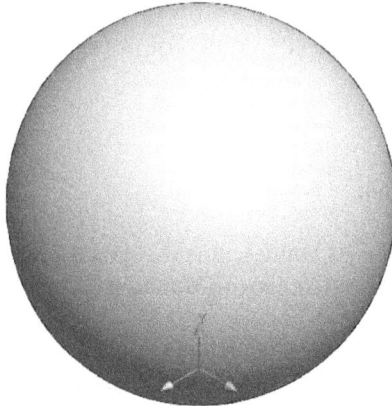

Figure 7-31 Solid model for Exercise 1

Exercise 2

In this exercise, you will open the model (*c04_exr02*) created in Exercise 2 of Chapter 4. You can also download this file from *www.cadcim.com* by using the following path:

Textbooks > CAE Simulation > Dassault Systemes > SOLIDWORKS Flow Simulation > Flow Simulation Using SOLIDWORKS 2023 > Input Files > C07_SWFS_inp

You will then apply goals to the model. The model is shown in Figure 7-32.

(Expected time: 20 min)

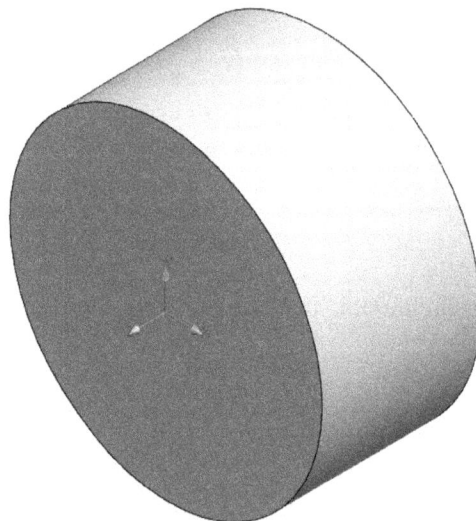

Figure 7-32 Solid model for Exercise 2

Answers to Self-Evaluation Test
1. Total Solution, 2. Global, 3. Name Template, 4. Point Goal, 5. Surface Source

Chapter 8

Analyzing Results

Learning Objectives

After completing this chapter, you will be able to:

- *Run the analysis*
- *Create cut plots*
- *Create surface plots*
- *Create flow trajectories*
- *Create goal plot*

INTRODUCTION

After adding goals to a project, you need to run the analysis and plot results to understand them. In this chapter, you will learn more about the tools used for performing an analysis and plot the results. These tools are discussed next.

RUN

CommandManager:	Flow Simulation > Run
SOLIDWORKS Menus:	Tools > Flow Simulation > Solve > Run
Toolbar:	Flow Simulation Main > Run

This tool is used to run the calculation process for a project. Choose the **Run** tool from the **Flow Simulation CommandManager**; the **Run** dialog box is displayed, as shown in Figure 8-1. The options in this dialog box are discussed next.

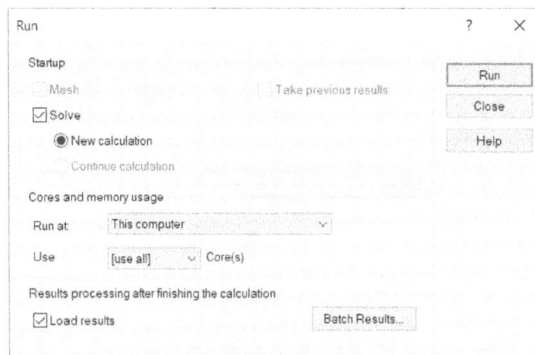

*Figure 8-1 The **Run** dialog box*

Startup Area

The options in the **Startup** area are discussed next.

Mesh

Select this check box if you want to create a new computational mesh for the project that is already meshed. But if the mesh is not calculated before then this check box will remain selected.

Solve

Select this check box to specify the calculation process for the project. You can select one of the following options:

New calculation

Select this radio button to recalculate the project if you have previously calculated this project by using the same settings as specified on the wizard page.

Continue calculation

Select this radio button to continue the calculation process that was terminated automatically or manually.

Take previous results

Select this check box to ignore the initial conditions specified in the wizard page while calculating the project.

Cores and memory usage Area

The options in the **Cores and memory usage** area are discussed next.

Run at

This drop-down is used to specify the option for running the calculation. The **This computer** option helps the solver to run on the current computer as a separate process. The **Add computer** option is used for adding the network computer for using its CPU and memory. As you select the **Add computer** option, the **Add computer** dialog box is displayed. In the **Name** edit box, you can enter the name or IP address of the network computer on which calculation will be run. In the **Port** edit box, you need to enter the port number of the network computer. Choose the **Add** button after specifying the name and port number. The **Browse** button can be used to add the available network computers. The added network computer will be displayed in the list with more details like number of available processors or cores, the amount of available memory and the type of operating system. Choose the **OK** button to close the dialog box.

Use

This drop-down helps you to select the number of cores for the calculation. The number of available cores depends upon the operating system or core installed in the computer.

Results Processing after Finishing the Calculation Area

The options in this area are discussed next.

Load results

When this check box is selected, the results are automatically loaded after the calculation is finished.

Batch Results

This button helps you to automatically create the standard results, goal plots and X-Y plots for the project. Choose the **Batch Results** button; the **Batch Results Processing** dialog box is displayed. From this dialog box, you can select the project from the list of projects for which you want to create plots and reports. You can choose the **Run** button to start batch result processing for the currently active project from the list. The **Run All** button is used to start the batch result processing for all the projects which are mentioned in the list.

Choose the **Run** button after specifying the options in the dialog box; the **Solver** window is displayed in which the mesh generation and solver calculation will take place, refer to Figure 8-2.

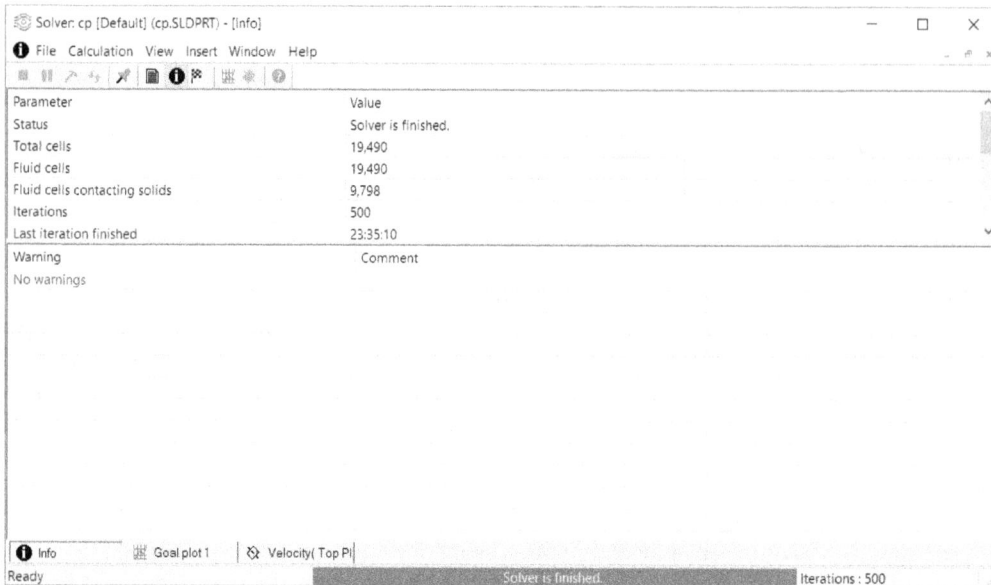

*Figure 8-2 The **Solver** window*

Also, the **Solver is finished** message is displayed at the bottom of the window, implying the calculation has been done.

CUT PLOT

CommandManager:	Flow Simulation > Results > Insert > Cut Plot
SOLIDWORKS Menus:	Tools > Flow Simulation > Results > Insert > Cut Plot

This tool helps you to view the section view of a parameter distribution. Choose the **Cut Plot** tool from the **Insert** drop-down in the **Flow Simulation CommandManager**; the **Cut Plot PropertyManager** is displayed, as shown in Figure 8-3. There are five rollouts in this property manager, namely **Selection**, **Display**, **Contours** (It is displayed by default.), **Options** and **Crop Regions**. These rollouts are discussed next.

Selection
This rollout is used to specify the section plane in which you can see the result. There are three buttons available in this rollout to specify the plane and they are discussed next.

Reference
By using this button, you can create a plane parallel to the selected plane, face, or normal to the curve.

XYZ Planes
By using this button, you can create a plane parallel to the default plane.

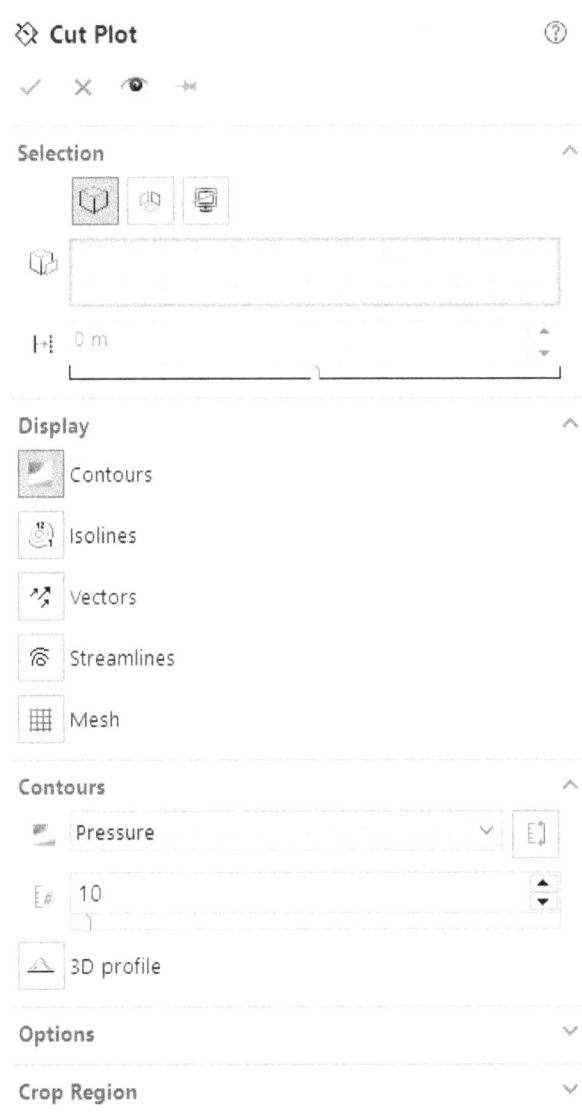

*Figure 8-3   The **Cut Plot PropertyManager***

Normal to Screen
By using this button, you can create a plane normal to the screen.

Display
This rollout is used to display the results depending on the selection of the option.

Contours
By using this button, you can display the distribution of parameter on a plane.

Isolines
By using this button, you can display the isoline of a parameter.

Vectors
By using this button, you can display vectors to visualize vector parameters.

Streamlines
By using this button, you can display field lines used to visualize vector parameters.

Mesh
By using this button, you can display the computational mesh on a plane.

Contours
This rollout is displayed when you choose the **Contours** button from the **Display** rollout. You can select the required parameter from the **Parameter** drop-down to display contours. Choose the **Adjust Minimum and Maximum** button to view the changes occurring with in the specific range. As you choose this button, the **Maximum** and **Minimum** edit boxes will be available to specify the values. You can specify the number of divisions into which the parameter range is divided in the **Number of Levels** edit box. Choose the **3D profile** button to view the parameter distribution as a 3D profile.

Isolines
This rollout is displayed when you choose the **Isoline** button from the **Display** rollout. The **Parameter** and **Number of Levels** options have already been discussed in the previous head. You can specify the width of the isolines in the **Width** edit box. You can select the option from the **Color by** drop-down to color the isolines. The **Display values** button is used to display or hide the parameter value on the isoline.

Vectors
This rollout is displayed when you choose the **Vectors** button from the **Display** rollout. You can either choose the **Static Vectors** or **Dynamic Vectors** button to specify the way vectors will be displayed. You can select the required parameter from the **Parameter** drop-down whose distribution you want to visualize with vector representation. The **Adjust Minimum and Maximum** button has been discussed earlier. The **Spacing** slider is used to control the distance between the vector starting points. The **Color by** drop-down is used to specify the color for the vectors.The **3D vectors** button is used to visualize vectors in three dimensions. The **Gradient plot** button is used to select the distribution of vectors.

Mesh
This rollout is displayed when you choose the **Mesh** button from the **Display** rollout. The **Color by** drop-down in this rollout allows you to display the fixed color background for the cut plot. Choose the **Color** button to specify the type of color.

Options
The **Use CAD geometry** check box in this rollout is used to display the original model by default when displaying results. The geometry utilized in the calculation may change slightly from the

original model geometry, depending on how precisely the model is resolved by the computational mesh. You need to clear this check box to use the flow simulation interpreted geometry.

The **Interpolate** check box is used to display the parameter distribution, with values obtained in the cell centres interpolated by the computational mesh cells.

The **Display outlines** check box is used to display or hide the plot area outlines. The Edge Color drop-down allows you to change the default outline color.

The **Display boundary layer** check box is used to display or hide boundary layers in the cut plot. It requires additional computer resources to visualize the boundary layer. You need to clear this check box for faster creation and modification of the cut plot.

Crop Region

The **Crop Region** check box is used to define a 3D box that crops the cut plot area, displaying only the portion of the plot that is inside the box. The **Xmax**, **Xmin**, **Ymax**, **Ymin**, **Zmax**, and **Zmin** edit boxes are available in this rollout to specify the 3D box values.

Choose the **OK** button after specifying the settings in the dialog box to close it.

SURFACE PLOT

CommandManager:	Flow Simulation > Results > Insert > Surface Plot
SOLIDWORKS menus:	Tools > Flow Simulation > Results > Insert > Surface Plot

This tool helps you to view the parameter distribution on the selected model faces or surfaces. Choose the **Surface Plot** tool from the **Insert** drop-down in the **Flow Simulation CommandManager**; the **Surface Plot PropertyManager** is displayed, as shown in Figure 8-4. There are five rollouts in this property manager, **Selection**, **Display**, **Contours** (It is displayed by default.), **Options**, and **Crop Regions**. All of them have been discussed in the previous section.

FLOW TRAJECTORIES

CommandManager:	Flow Simulation > Results > Insert > Flow Trajectories
SOLIDWORKS menus:	Tools > Flow Simulation > Results > Insert > Flow Trajectories

This tool helps you to view the flow trajectories as flow streamlines. Choose the **Flow Trajectories** tool from the **Insert** drop-down in the **Flow Simulation CommandManager**; the **Flow Trajectories PropertyManager** is displayed, as shown in Figure 8-5. There are four rollouts in this property manager: **Starting Points**, **Appearance**, **Constraints** and **Crop Region**. They are discussed next.

◇ Surface Plot   ⑦   ▦ Flow Trajectories   ⑦

✓ ✕ ⤴   ✓ ✕ ◉ ⤴

Selection   ⌃   Starting Points   ⌃

⬡ [   ]   :: ⬚ ⌖ ⬚ ˣ_z ⬚ ⬚

                                             ⬡ [   ]

□ Use all faces   □ Mesh based point generation

Display   ⌃   ⤳ [20   ] ▲▼

🖼 Contours   ⤳ [0.003 m   ] ▲▼

🖼 Isolines   Appearance   ⌃

⤳ Vectors   ⤳ ⬚

🖼 Streamlines   ⤳ [Arrows   ⌄]

▦ Mesh   ▷ [0.0009 m   ] ▲▼

Contours   ⌃   🖼 [Pressure   ⌄] 🖼

🖼 [Pressure   ⌄] 🖼   ⤳ [0   ] ▲▼

Ɇ# [100   ] ▲▼   Constraints   ⌄

Options   ⌄   Crop Region   ⌄

Crop Region   ⌄

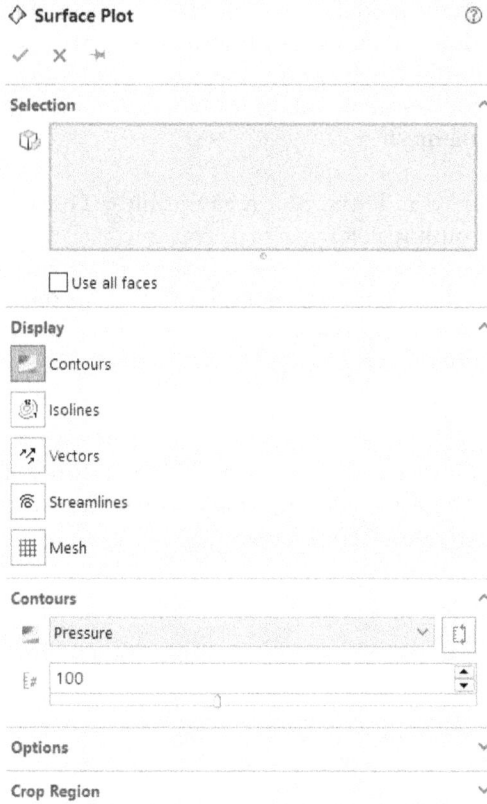

Figure 8-4 *The Surface Plot PropertyManager*   **Figure 8-5**   *The Flow Trajectories PropertyManager*

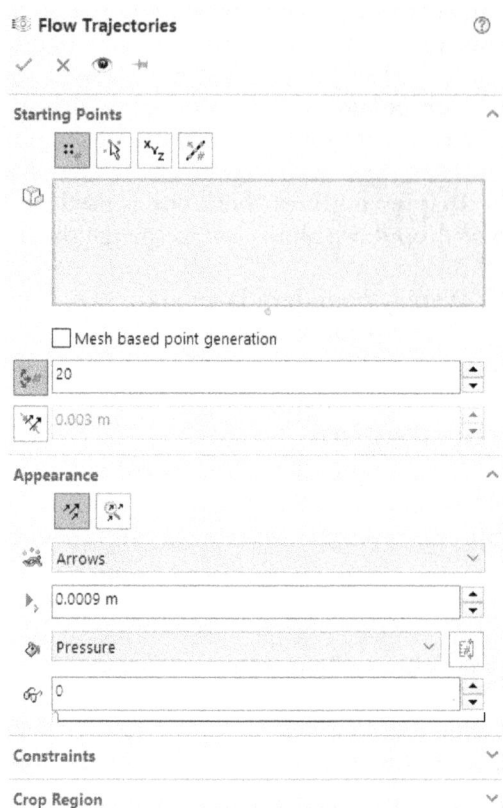

Starting Points

This rollout is used to define the starting points of the trajectories. Four buttons are available in this rollout and they are discussed next.

Pattern

By using this button, the starting points of the trajectories are evenly divided over the specified planes, faces, sketches, edges, and curves. Select a face from the drawing area; the selected faces will be displayed in the **Planes**, **Faces**, **Sketches**, **Edges**, and **Curves** areas.

Select the **Mesh based point generation** check box to distribute the starting points throughout the computational mesh cells on the selected planes, model faces, surfaces, drawings, curves, or edges.

Select the **In plane** check box to get flow trajectories on a plane or planar surface by integrating in-plane components of velocity.

In the **Number of Points** edit box, you can specify the number of points which you want to display and in the **Spacing** edit box, you can enter the value for spacing between the starting point of trajectories.

Pick from Screen

By using this button, you can select starting points on a plane or planar face in the graphics area. Select the face from the drawing area, the selected faces will be displayed in the **Section Plane, Planar Faces, or Curve** area. You can offset the starting points plane from the selected plane or face by using the slider or input the offset value in the **Offset** edit box. The **Pick points** button is used to the select the point from the drawing area. You can specify a set of starting points by specifying points in the drawing area whose x,y, and z coordinates are visible in the **Coordinates in Global Coordinates System** area. The **Delete Point** button is used to delete the selected point from the **Coordinates in Global Coordinates System** area and the **Delete All** button is used to delete all the points in the **Coordinates in Global Coordinates System** area.

Coordinates

By using this button, you can create a table of point coordinates which allows you to specify a set of starting points. Enter the coordinate value of the point in the **X**, **Y**, and **Z Coordinate** edit boxes and then click on the **Add point** button to add this point in the **Coordinates in Global Coordinates System** area. You can edit a coordinate value of the point by double-clicking on the coordinate value in the **Coordinates in Global Coordinates System** area. The **Delete Point** button is used to delete the selected point and the **Delete All** button is used to delete all the points in the **Coordinates in Global Coordinates System** area.

Pattern on Shapes

By using this button, you can create trajectory starting points that are evenly distributed over a line, rectangle or sphere. You can select the **Line**, **Rectangle** or **Sphere** radio button from the **Starting Points** rollout to distribute the starting points. Select the **Line** radio button if you want that the starting points to be distributed over a line. The default line size and position are specified automatically. Select the **Rectangle** radio button to evenly distribute the starting points rectangularly. The default rectangle size and position are specified automatically. Select the **Sphere** radio button to evenly distribute the starting points spherically. The default sphere size and position are specified automatically.

Appearance

This rollout is used to define the display settings for trajectories. There are two buttons, **Static Trajectories,** and **Dynamic Trajectories** available in this rollout. They are used to control the display settings of trajectories and are discussed next.

Static Trajectories

By using this button, you can display the trajectories as static image. When you choose this button, the **Pipes, Lines, Lines with Arrows, Bands, Spheres, Arrows,** and **Arrows (flat)** options are displayed in the **Draw Trajectories As** drop-down list. In the **Arrow Size** edit box, you can enter the size of an arrow which shows the trajectories path. In the **Color by** drop-down list, you can select the option to determine the path to color the flow trajectories. The **Plot Transparency** edit box or slider is used to adjust the transparencies of trajectories in the range of 0 to 1.

Dynamic Trajectories

By using this button, you can display the trajectories in real time. When you choose this button, the **Arrows** and **Spheres** options are displayed in the **Draw Trajectories As** drop-down list. In the **Size** edit box, you can enter the size of an arrow or sphere depending upon the selection of option from the **Draw Trajectories As** drop-down list.

Constraints

This rollout is used to apply constraints to the trajectories. There are **Forward**, **Backward**, and **Both** buttons available in this rollout. These buttons are used to specify the direction of flow trajectories relative to the starting point of the trajectories. If you click the **Forward** button, the streamline is displayed as a starting point from this point. If you click the **Backward** button, the streamline is displayed as ending point at this point. If you select the **Both** button, it will display a whole flow streamline passing through the selected point, from start to end. You can enter the value in the **Maximum Length** edit box to limit the length of the trajectory. In the **Maximum Time** edit box, enter the value to stop the trajectory when the flow traveling time has reached the specified maximum value. By default, the **Use CAD geometry** check box is selected which implies that flow simulation uses the original model while creating trajectories.

The **Crop Region** rollout is already discussed previously.

GOAL PLOT

CommandManager:	Flow Simulation > Insert > Goal Plot
SOLIDWORKS menus:	Tools > Flow Simulation > Results > Insert > Goal Plot
Toolbar:	Flow Simulation Results Insert Group > Goal Plot

This tool is used to track goal changes as the calculation proceeds. Choose the **Goal Plot** tool from the **Insert** drop-down in the **Flow Simulation CommandManager**; the **Goal Plot PropertyManager** will be displayed, as shown in Figure 8-6. You can select the type of goal from the **Goal Filter** drop-down list to display the selected goals in the **Goals to Plot** area of the **Goals** rollout. Select the goals which you want to display by selecting the check boxes at the left of each goal's name in the **Goals to Plot** area. Select the **Group charts by parameter** check box from the **Options** rollout to plot all goals for the same parameter on the same chart. Click the **Show** button to display the calculated parameter values at the bottom of the screen. The **Export to Excel** button is used to calculate the parameters and export the calculated data into Microsoft Excel.

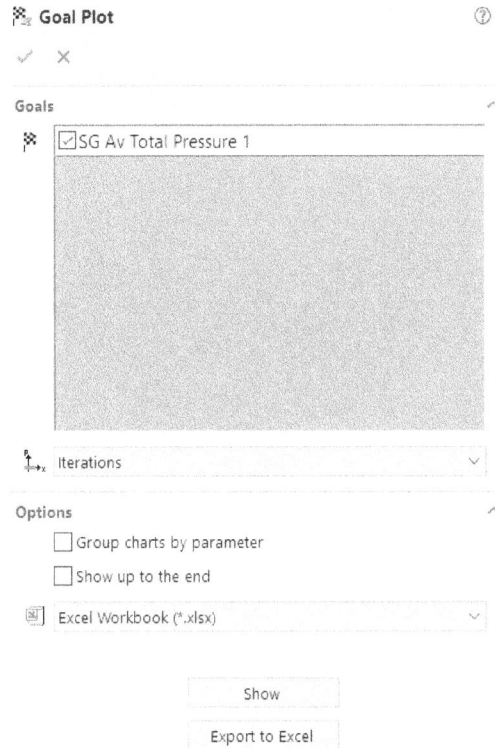

*Figure 8-6   The **Goal Plot PropertyManager***

TUTORIALS

Tutorial 1

In this tutorial, you will open the model (*c07_tut01*) created in Tutorial 1 of Chapter 7. You can also download this file from *www.cadcim.com* by using the following path:

Textbooks > CAE Simulation > Dassault Systemes > SOLIDWORKS Flow Simulation > Flow Simulation Using SOLIDWORKS 2023 > Input Files > C08_SWFS_inp

You will then visualize the results of the simulation. The model is shown in Figure 8-7.

(Expected time: 20 min)

The following steps are required to complete this tutorial:

a. Open Tutorial 1 of Chapter 7.
b. Save this tutorial in the *c08* folder with a new name.
c. Add cut plots to the model.
d. Add flow trajectories to the model.

e.   Add goal plots to the model.
f.   Save the file.

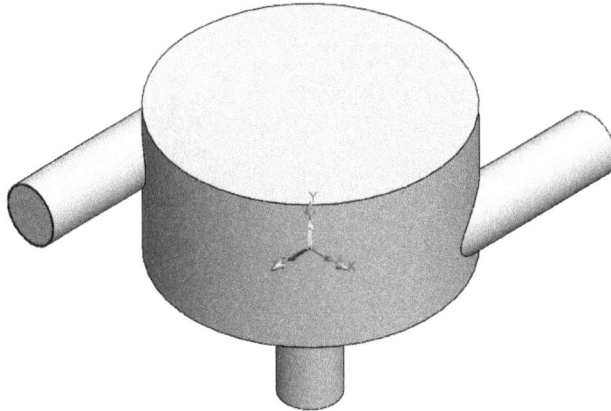

Figure 8-7   *Model for Tutorial 1*

Opening Tutorial 1 of Chapter 7

As the required tutorial is saved in the *c07* folder, you need to select this folder and then open the *c07_tut01.sldprt* document.

1.   Start SOLIDWORKS by double-clicking on its shortcut icon on the desktop of your computer.

2.   Choose the **Open** button from the Menu Bar to display the **Open** dialog box, refer to Figure 8-8.

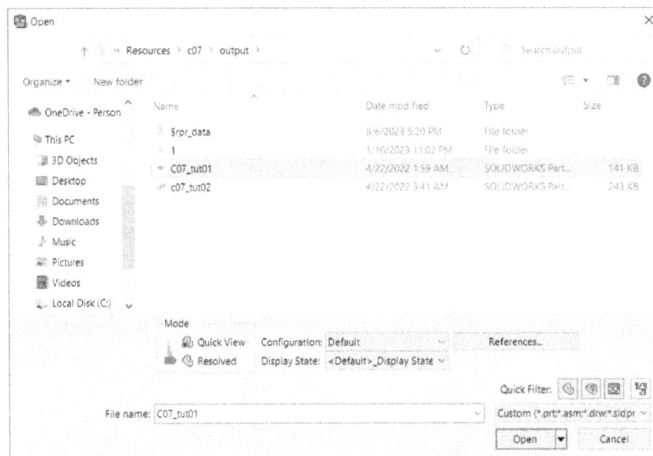

Figure 8-8   *The **Open** dialog box*

3.   Browse to the SOLIDWORKS folder and select the *c07* folder.

4.   Select the **c07_tut01.sldprt** document and then choose the **Open** button.

As the model was saved in the Flow Simulation environment in Chapter 7, it opens in the Flow Simulation environment, refer to Figure 8-9.

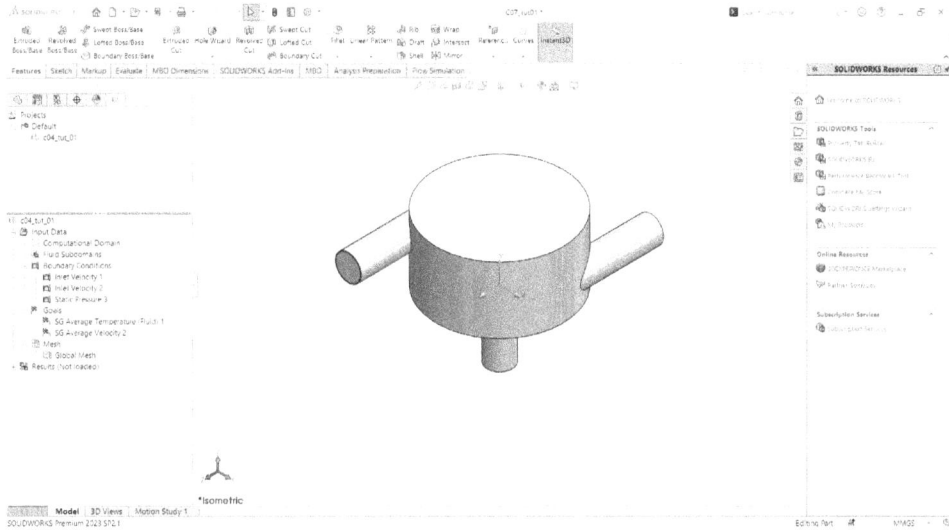

Figure 8-9 *The model opened in the Flow Simulation environment*

Saving the Document in the c08 Folder

When you open a document from another chapter, it is recommended that you first save the opened document with a new name in the folder of the current chapter to avoid the original document from getting modified.

1. Choose the **Save As** button from the **Save** flyout in the Menu Bar; the **Save As** dialog box is displayed, refer to Figure 8-10.

2. Browse to the **SOLIDWORKS_Flow > Resources** folder and then create a new folder with the name **c08** by using the **Create New Folder** button. Make the **c08** folder as the current folder by double-clicking on it.

3. Enter **c08_tut01** as the new name of the document in the **File name** edit box and then choose the **Save** button to save the document. The document is saved with the new name and gets opened in the drawing area.

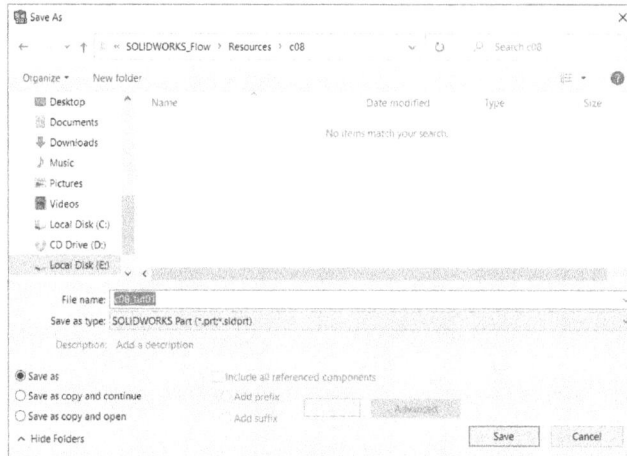

Figure 8-10   *The Save As dialog box*

Running the Calculation

Next, you need to invoke the **Run** tool and start the calculation.

1. Choose the **Run** tool from the **Flow Simulation CommandManager**; the **Run** dialog box is displayed, as shown in Figure 8-11.

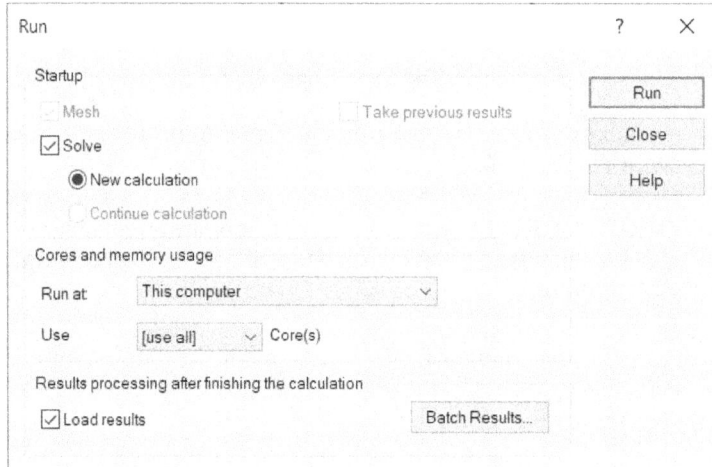

Figure 8-11   *The Run dialog box*

2. Select the **Mesh** check box and also ensure that the **Solve** check box is selected. Choose the **Run** button; first the **Geometry Preparation**, and then **Mesh Generation**, and at last the **Solver** windows are displayed. Note that window name change as the respective operation finishes. Finally, the **Solver** window is displayed and the "**Solver is finished**" message box is displayed.

3. Now, choose the **Close** button to close the dialog box.

4. Right-click on the **Results** node from the **Flow Simulation Analysis Tree** and select the option **load** to load results.

Creating the Cut Plot

Next, you need to invoke the **Cut Plot** tool and observe the velocity plot.

1. Choose the **Cut Plot** tool from the **Insert** drop-down in the **Flow Simulation CommandManager**; the **Cut Plot PropertyManager** is displayed, as shown in Figure 8-12.

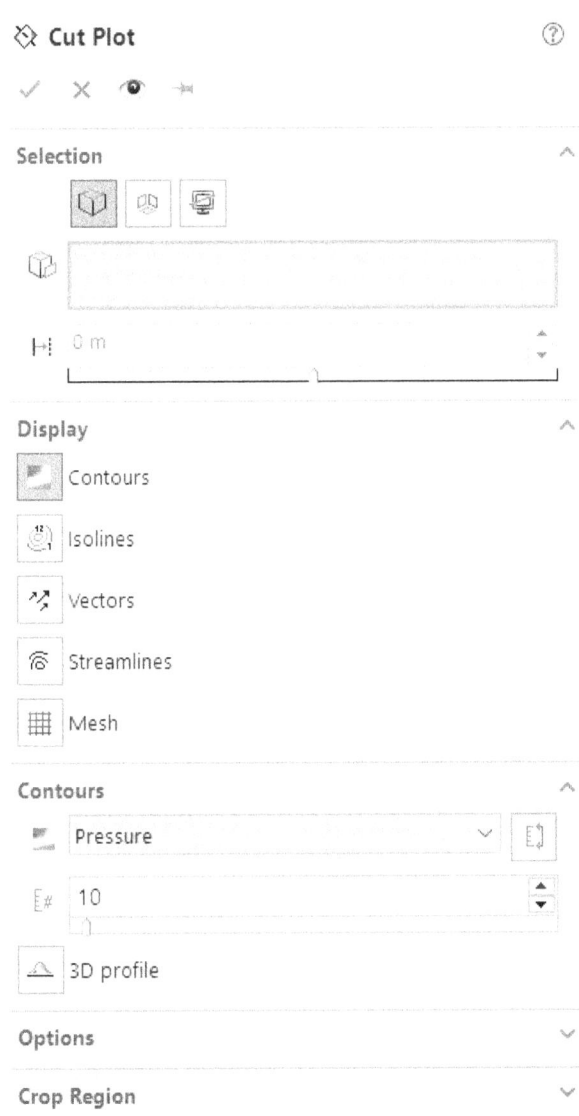

*Figure 8-12 The **Cut Plot PropertyManager***

2. Select the Front Plane from the drawing area which is displayed in the **Section Plane**, **Planar Face** or **Curve Selection** box in the **Selection** rollout. Ensure that the **Reference** button is chosen in the **Selection** rollout.

3. Select the **Contours** button from the **Display** rollout if it is not selected.

4. Select the **Velocity** parameter from the **Parameter** drop-down list in the **Contours** rollout.

5. Choose the **OK** button to close the dialog box; the **Cut Plot** is displayed in the drawing area, refer to Figure 8-13. To view the cut plot, hide the model. Here you will notice that the maximum velocity is 11.77 m/s.

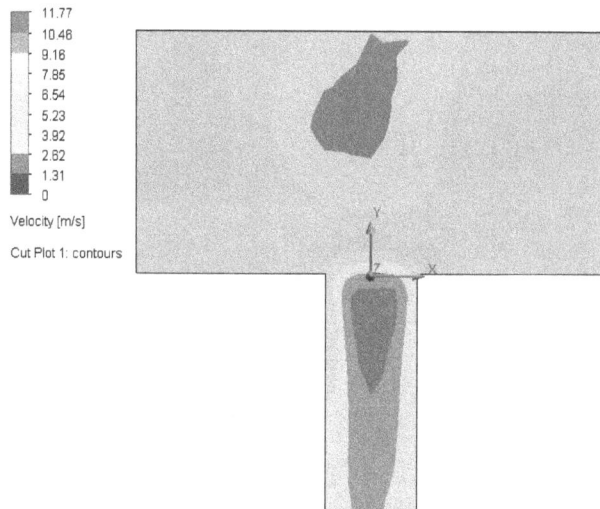

Figure 8-13 The Velocity cut plot

Creating the Flow Trajectories

Next, you need to invoke the **cut plot** tool and observe the velocity plot.

1. Choose the **Flow Trajectories** tool from the **Insert** drop-down in the **Flow Simulation CommandManager**; the **Flow Trajectories PropertyManager** is displayed, as shown in Figure 8-14.

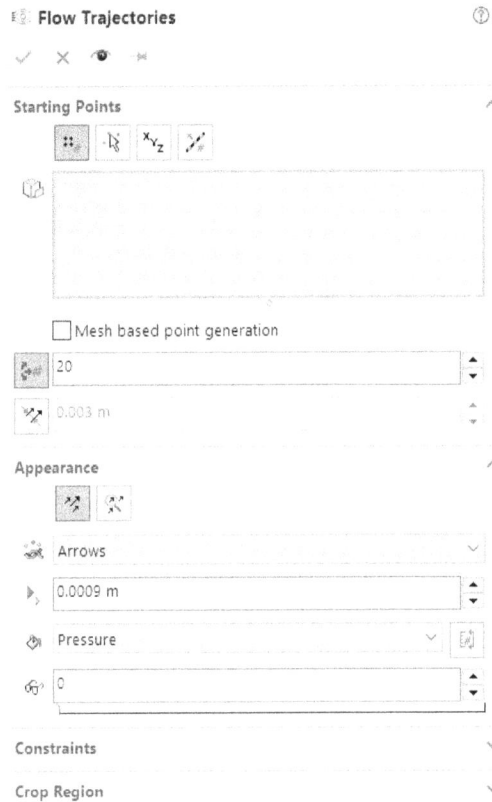

Figure 8-14 The Flow Trajectories PropertyManager

2. Select the Front Plane from the drawing area which is displayed in the **Planes, Faces, Sketches, Edges and Curves** selection box in the **Starting Points** rollout. Ensure that the **Pattern** button is selected in the **Starting Points** rollout.

3. Select the **Lines with Arrows** option from the **Draw Trajectories As** drop-down list from the **Appearance** rollout. Ensure that the **Static Trajectories** button is chosen in the **Appearance** rollout.

4. Select the **Temperature(Fluid)** option from the **Color by** drop-down list in the **Appearance** rollout.

5. Choose the **OK** button to close the dialog box; the trajectories are displayed in the drawing area. To view the trajectories, hide the model. You will notice the temperature distribution in the model, refer to Figure 8-15.

Figure 8-15 *The Temperature(Fluid) flow trajectories*

Creating the Goal Plot

Next, you need to invoke the **Goal Plot** tool and observe the velocity plot.

1. Choose the **Goal Plot** tool from the **Insert** drop-down in the **Flow Simulation CommandManager**; the **Goal Plot PropertyManager** is displayed, as shown in Figure 8-16.

2. Select the **SG Average Temperature (Fluid) 1** check box from the **Goals to Plot** list in the **Goals** rollout.

3. Choose the **OK** button to close the PropertyManager; the Temperature versus Iterations graph is displayed in the drawing area, refer to Figure 8-17.

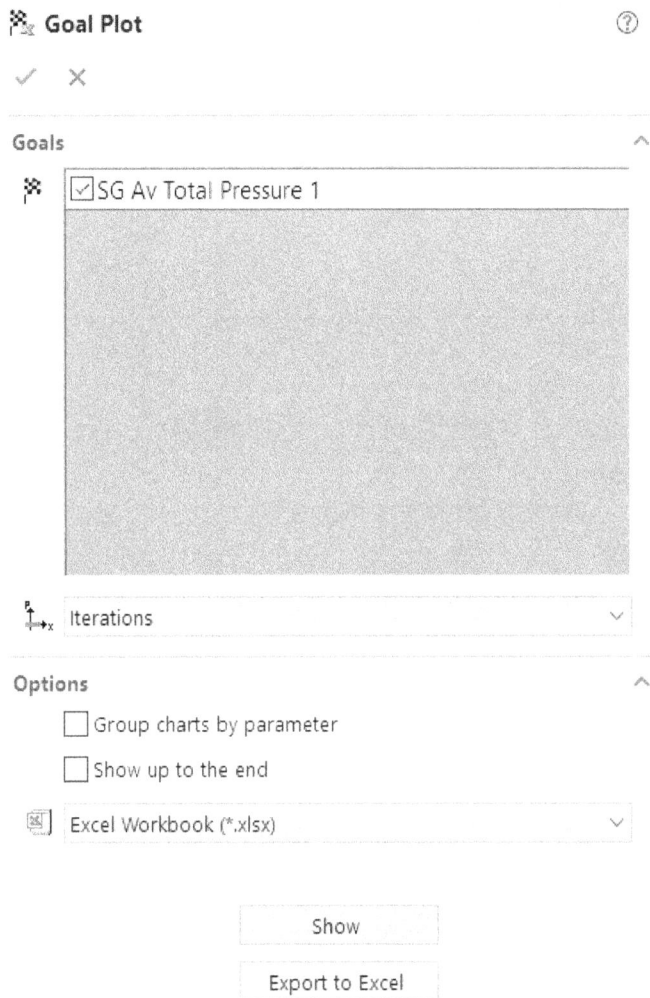

Figure 8-16   The Goal Plot PropertyManager

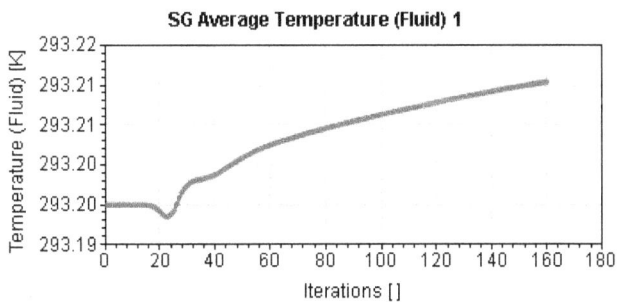

Figure 8-17   The Temperature(Fluid) goal plot

4. Follow steps 1 to 3 and create a Velocity vs Iterations graph in the drawing area, refer to Figure 8-18.

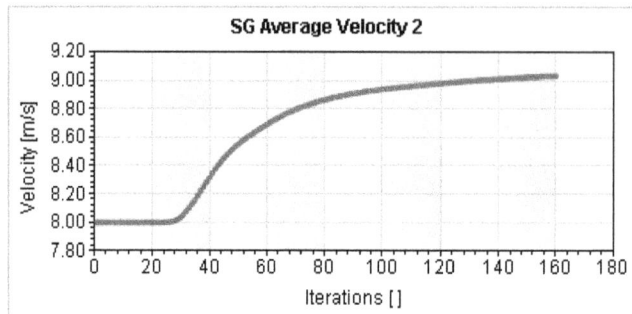

Figure 8-18 *The Velocity goal plot*

Saving the Model

1. Save the part document with the name **c08_tut01** at the following location: *SOLIDWORKS_Flow\ Resources\c08*.

2. Choose **File > Close** from the SOLIDWORKS menus to close the document.

Tutorial 2

In this tutorial, you will open the model (*c07_tut02*) created in Tutorial 2 of Chapter 7. You can also download this file from *www.cadcim.com* by using the following path:

Textbooks > CAE Simulation > Dassault Systemes > SOLIDWORKS Flow Simulation > Flow Simulation Using SOLIDWORKS 2023 > Input Files > C08_SWFS_inp

You will then visualize the results of the simulation. The model is shown in Figure 8-19.

(Expected time: 20 min)

The following steps are required to complete this tutorial:

a. Open Tutorial 2 of Chapter 7.
b. Save this tutorial in the *c08* folder with a new name.
c. Add cut plots to the model.
d. Add flow trajectories to the model.
e. Add goal plots to the model.
f. Save the file.

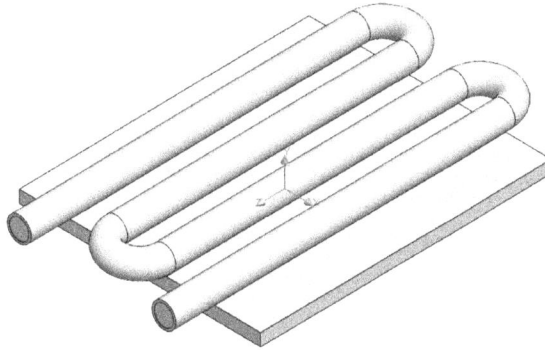

Figure 8-19 Model for Tutorial 2

Opening Tutorial 2 of Chapter 7

As the required document is saved in the *c07* folder, you need to select this folder and then open the *c07_tut02.sldprt* document.

1. Start SOLIDWORKS by double-clicking on its shortcut icon on the desktop of your computer.

2. Choose the **Open** button from the Menu Bar to display the **Open** dialog box, refer to Figure 8-20.

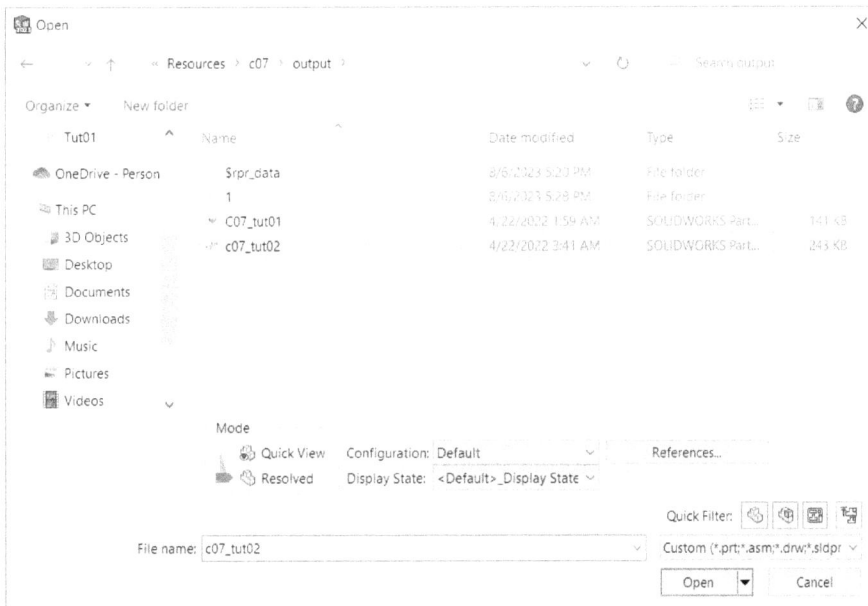

Figure 8-20 The **Open** dialog box

3. Browse to the SOLIDWORKS folder and select the **c07** folder.

4. Select the **c07_tut02.sldprt** document and then choose the **Open** button.

Saving the Document

When you open a document from another chapter, it is recommended that you first save the opened document with a new name in the folder of the current chapter to avoid the original document from getting modified.

1. Choose the **Save As** button from the **Save** flyout in the Menu Bar; the **Save As** dialog box is displayed, as shown in Figure 8-21.

2. Browse to the **SOLIDWORKS_Flow > Resources** folder and make the **c08** folder as the current folder by double-clicking on it.

3. Enter **c08_tut02** as the new name of the document in the **File name** edit box and then choose the **Save** button to save the document. The document is saved with the new name and gets opened in the drawing area.

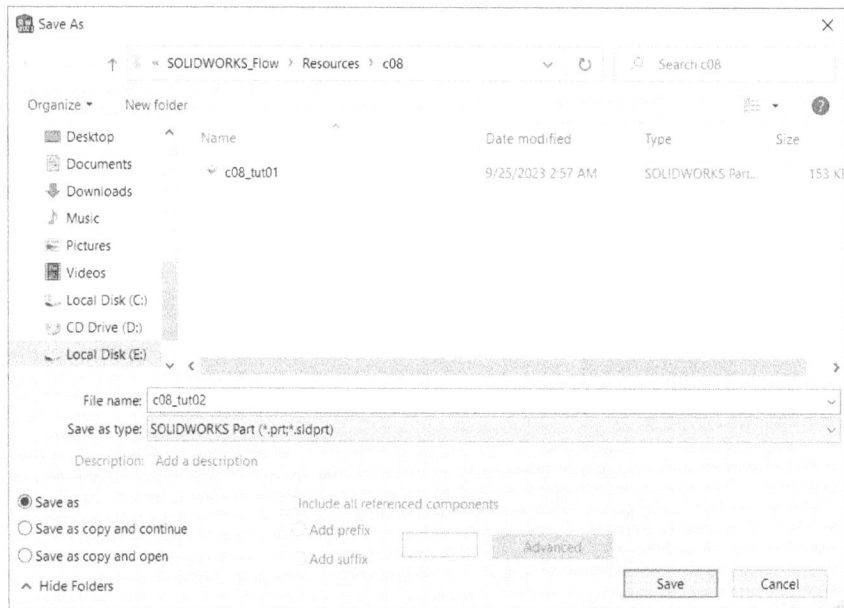

Figure 8-21   The ***Save As*** *dialog box*

Running the Calculation

Next, you need to invoke the **Run** tool and start calculation.

1. Choose the **Run** tool from the **Goals** drop-down in the **Flow Simulation CommandManager**; the **Run** dialog box is displayed, as shown in Figure 8-22.

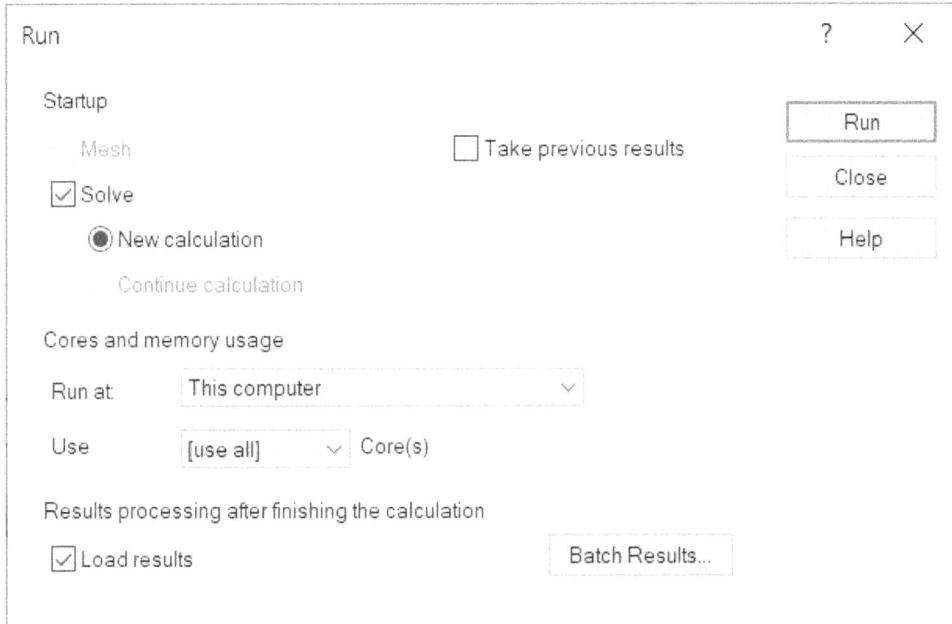

Figure 8-22   *The* ***Run*** *dialog box*

2. Select the **Mesh** check box and also ensure that the **Solve check** box is selected by default. Choose the **Run** button; first the **Geometry Preparation** and then the **Mesh Generation** and at last the **Solver** windows are displayed. Note that window name change as the respective operation finishes. Finally, the **Solver** dialog box is displayed and the "**Solver is finished**" message box is displayed.

3. Now, choose the **Close** button to close the dialog box.

4. Right-click on the **Results** node from the **Flow Simulation Analysis Tree** and select the **Load** option to load the results.

Creating the Cut Plot

Next, you need to invoke the **Cut Plot** tool and observe the velocity plot.

1. Choose the **Cut Plot** tool from the **Insert** drop-down in the **Flow Simulation CommandManager**; the **Cut Plot PropertyManager** is displayed, as shown in Figure 8-23.

2. Select the Top Plane from the drawing area which is displayed in the Section Plane, Planar Face or Curve selection box in the **Selection** rollout. Ensure that the **Reference** button is chosen in the **Selection** rollout. If you need to change the position of plane, use the green arrow from the drawing area.

3. Choose the **Contours** button from the **Display** rollout if it is not chosen.

4. Select the **Temperature (Fluid)** parameter from the **Parameter** drop-down list in the **Contours** rollout.

5. Choose the **OK** button to close the PropertyManager; the **Temperature (Fluid)** cut plot is displayed in the drawing area, refer to Figure 8-24. Note that, to view the cut plot, hide the model from the drawing area. Here, you will notice that the maximum temperature is 129.08 Celsius.

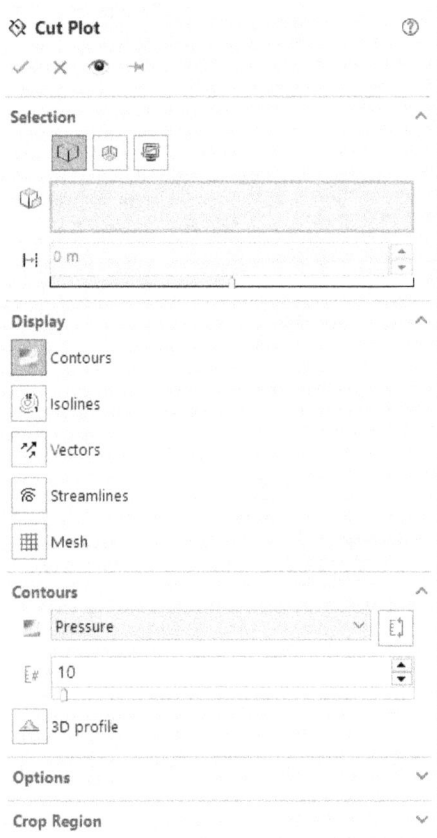

Figure 8-23 *The Cut Plot PropertyManager*

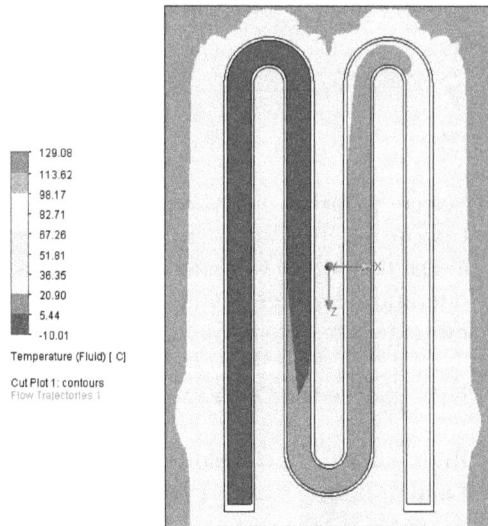

Figure 8-24 *The Temperature (Fluid) cut plot*

Creating the Flow Trajectories

Next, you need to invoke the **Cut Plot** tool and observe the velocity plot.

1. Choose the **Flow Trajectories** tool from the **Insert** drop-down in the **Flow Simulation CommandManager**; the **Flow Trajectories PropertyManager** is displayed, as shown in Figure 8-25.

2. Select the inside face of the Lid 1 from the drawing area which is displayed in the **Planes Faces**, **Sketches**, **Edges** and **Curves** selection box in the **Starting Points** rollout. Ensure that the **Pattern** button is chosen in the **Starting Points** rollout.

3. Select the **Lines with Arrows** option from the **Draw Trajectory As** drop-down list from the **Appearance** rollout. Ensure that the **Static Trajectories** button is chosen in the **Appearance** rollout.

4. Select the **Temperature (Fluid)** option from the **Color by** drop-down list in the appearance rollout.

5. Choose the **OK** button to close the PropertyManager; the trajectories are displayed in the drawing area. To view the trajectories, hide the model. You will notice the temperature distribution in the model, refer to Figure 8-26.

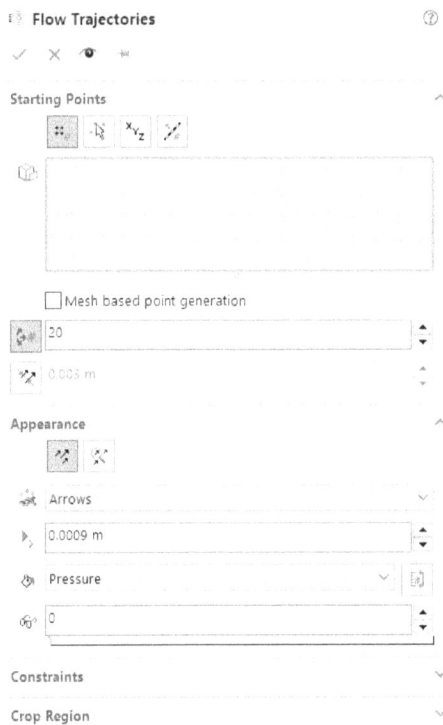

Figure 8-25 The Flow Trajectories PropertyManager

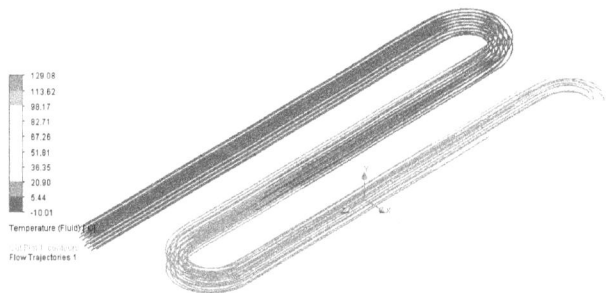

Figure 8-26 The Temperature (Fluid) flow trajectories

Creating the Goal Plot

Next, you need to invoke the **Goal Plot** tool and observe the velocity plot.

1. Choose the **Goal Plot** tool from the **Insert** drop-down in the **Flow Simulation CommandManager**; the **Goal Plot PropertyManager** is displayed, as shown in Figure 8-27.

2. Select all check boxes from the **Goals to Plot** list in the **Goals** rollout.

3. Choose the **Show** button from the **Options** rollout; the **Goal Plot 1** tab is added to the bottom of the drawing area. Here, you will note down the temperature values.

4. Choose the **OK** button to close the PropertyManager; the Temperature versus Iterations graph is displayed in the drawing area. You will notice that the maximum value of temperature of fluid is approximately 127 °C, refer to Figure 8-28.

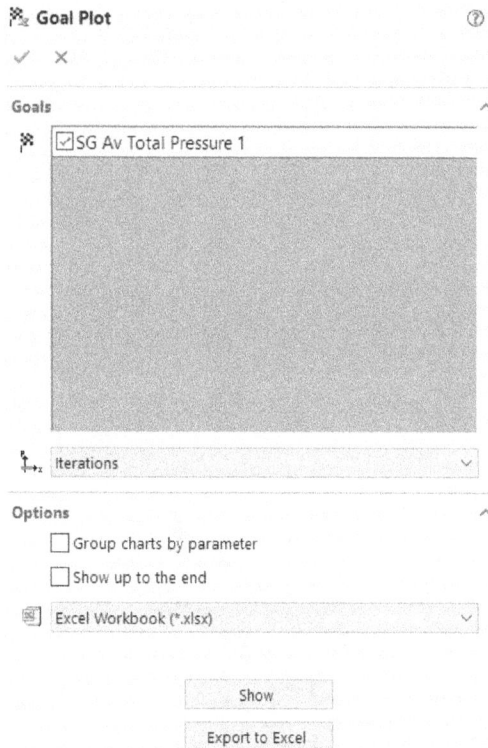

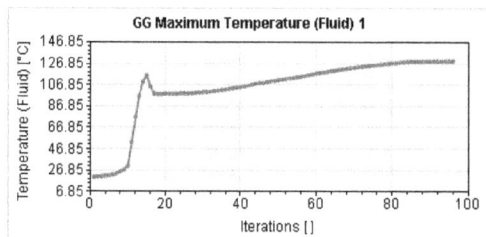

Figure 8-27 *The **Goal Plot PropertyManager***   *Figure 8-28* *The **Temperature (Fluid)** goal plot*

Creating the Surface Plot

Next, you need to invoke the **Surface Plot** tool and observe the velocity plot.

1. Choose the **Surface Plot** tool from the **Insert** drop-down in the **Flow Simulation CommandManager**; the **Surface Plot PropertyManager** is displayed, as shown in Figure 8-29.

2. Select the bottom face of the plate from the drawing area, the face is selected in the **Surfaces** area in the **Selection** rollout.

3. Choose the **Contours** button from the **Display** rollout if it is not chosen.

4. Select the **Temperature (Solid)** parameter from the **Parameter** drop-down list in the **Contours** rollout.

5. Choose the **OK** button to close the PropertyManager; the Temperature (Solid) graph is displayed in the drawing area. You will see that the maximum temperature of the solid is around 129 °C, refer to Figure 8-30.

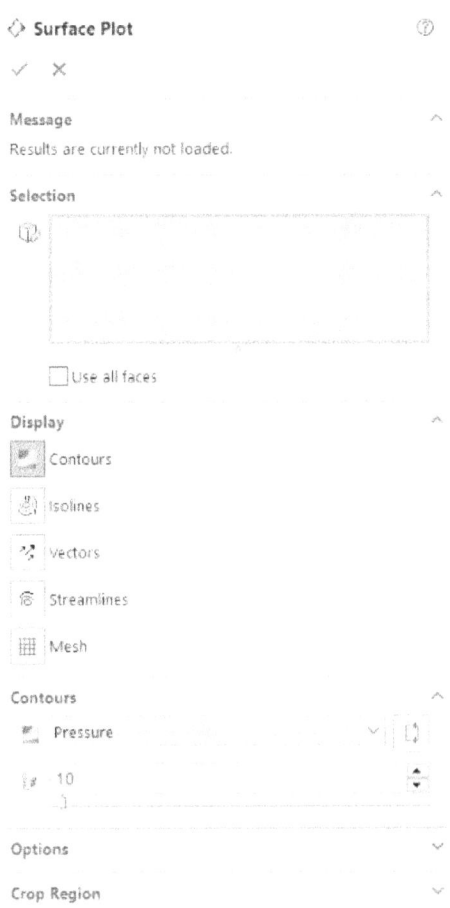

Figure 8-29 The **Surface Plot** PropertyManager

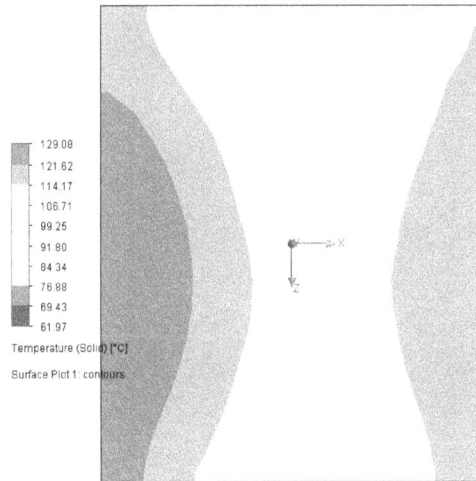

Figure 8-30 The solid surface plot

Saving the Model

1. Save the part document with the name *c08_tut02* at the following location: *\SOLIDWORKS_ Flow\resources\c08*.

2. Choose **File > Close** from the SOLIDWORKS menus to close the document.

Tutorial 3

In this tutorial, you will open the model (*c07_tut03*) created in Tutorial 3 of Chapter 7. You can also download this file from *www.cadcim.com* by using the following path:

Textbooks > CAE Simulation > Dassault Systemes > SOLIDWORKS Flow Simulation > Flow Simulation Using SOLIDWORKS 2023 > Input Files > C08_SWFS_inp

You will then visualize the results of the simulation. The model is shown in Figure 8-31.

(Expected time: 20 min)

The following steps are required to complete this tutorial:

a. Open Tutorial 3 of Chapter 7.
b. Save this tutorial in the *c08* folder with a new name.
c. Add cut plots to the model.
d. Add flow trajectories to the model.
e. Add goal plots to the model.
f. Save the file.

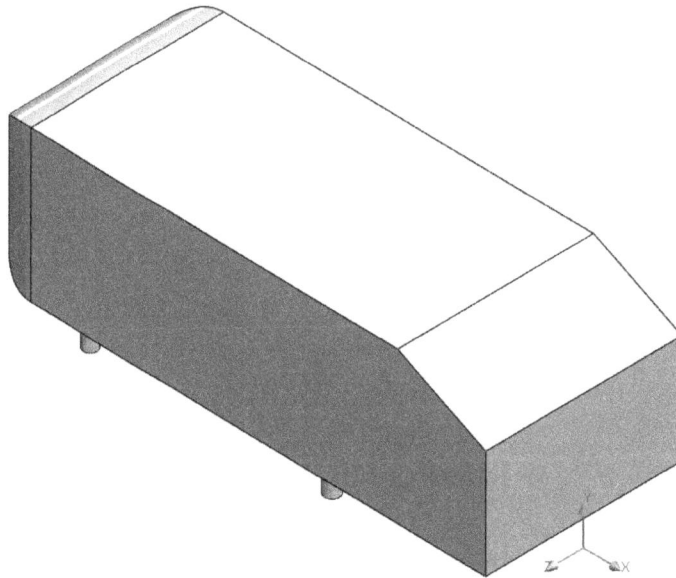

Figure 8-31 Model for Tutorial 3

Opening Tutorial 3 of Chapter 7

As the required document is saved in the *c07* folder, you need to select this folder and then open the *c07_tut03.sldprt* document.

1. Start SOLIDWORKS by double-clicking on its shortcut icon on the desktop of your computer.

2. Choose the **Open** button from the Menu Bar to display the **Open** dialog box, refer to Figure 8-32.

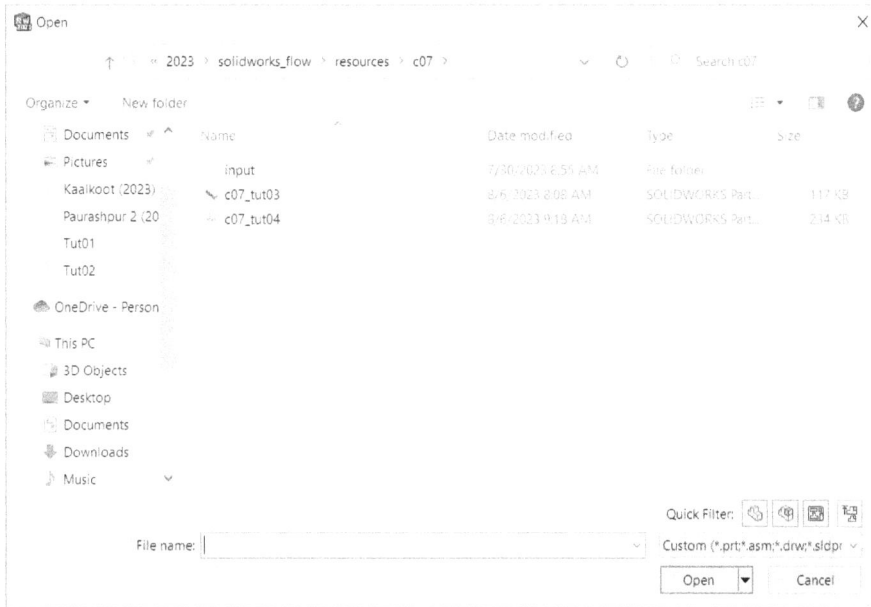

*Figure 8-32 The **Open** dialog box*

3. Browse to the SOLIDWORKS folder and select the **c07** folder.

4. Select the **c07_tut03.sldprt** document and then choose the **Open** button.

Saving the Document

When you open a document from another chapter, it is recommended that you first save the opened document with a new name in the folder of the current chapter to avoid the original document from getting modified.

1. Choose the **Save As** button from the **Save** flyout in the Menu Bar; the **Save As** dialog box is displayed, as shown in Figure 8-33.

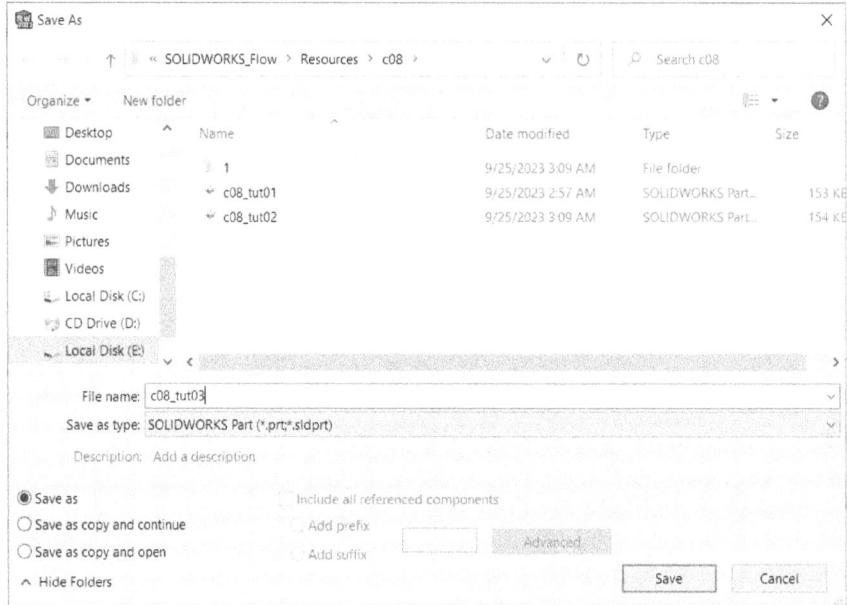

Figure 8-33   *The Save As dialog box*

2.   Browse to the **SOLIDWORKS_Flow > Resources** folder and make the *c08* folder as the current folder by double-clicking on it.

3.   Enter **c08_tut03** as the new name of the document in the **File name** edit box and then choose the **Save** button to save the document. The document is saved with the new name and gets opened in the drawing area.

Running the Calculation
Next, you need to invoke the **Run** tool and start the calculation.

1.   Choose the **Run** tool from the **Flow Simulation CommandManager**; the **Run** dialog box is displayed, as shown in Figure 8-34.

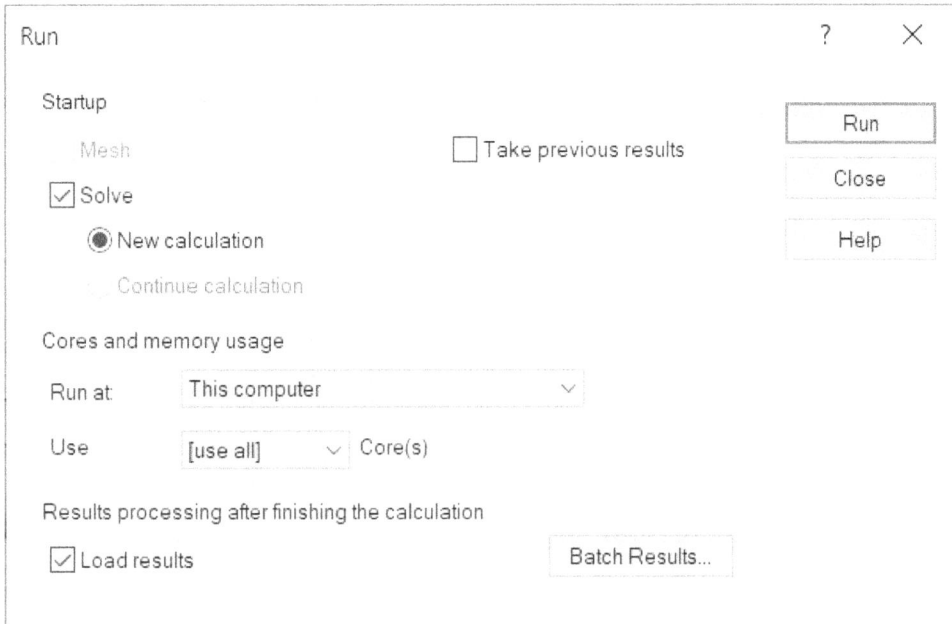

Figure 8-34   *The* **Run** *dialog box*

2.   Select the **Mesh** check box and also ensure that the **Solve** check box is selected by default. Choose the **Run** button; first the **Geometry Preparation**, then the **Mesh Generation** and then the **Solver** windows are displayed. Note that these window names change as the respective operation finishes. Finally, the **Solver** dialog box is displayed and the "**Solver is finished**" message box is displayed.

3.   Now, choose the **Close** button to close the dialog box.

4.   Right-click on the **Results** node from the **Flow Simulation Analysis Tree** and select the **Load** option to load the results.

Creating the Cut Plot

Next, you need to invoke the Cut Plot tool and observe the velocity plot.

1.   Choose the **Cut Plot** tool from the **Insert** drop-down in the **Flow Simulation CommandManager**; the **Cut Plot PropertyManager** is displayed, as shown in Figure 8-35. If this tool not available from the drop-down then select it from the tree under the Results node.

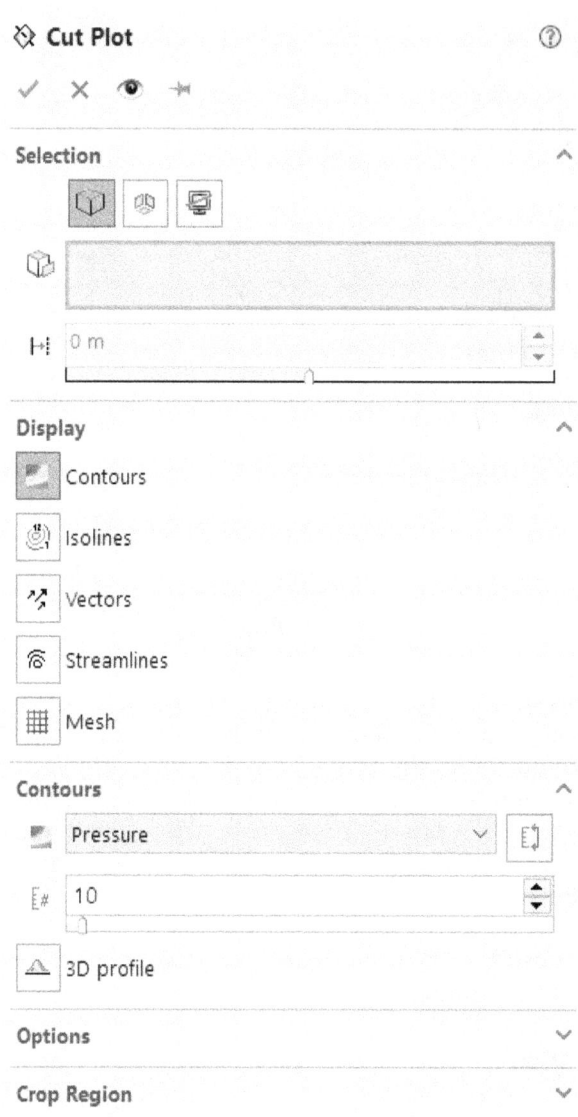

Figure 8-35   The Cut Plot PropertyManager

2. Select the Front Plane from the drawing area which is displayed in the **Section Plane, Planar Face or Curve selection** box in the **Selection** rollout.

3. Choose the **Contours** button from the **Display** rollout if it is not chosen.

4. Select the **Velocity (X)** parameter from the **Parameter** drop-down list in the **Contours** rollout.

5. Choose the **OK** button to close the PropertyManager; the **Velocity (X)** cut plot is displayed in the drawing area, refer to Figure 8-36. Note that, to view the cut plot, hide the model from the drawing area. Here, you will notice that the maximum Velocity(X) is 45.85 m/s.

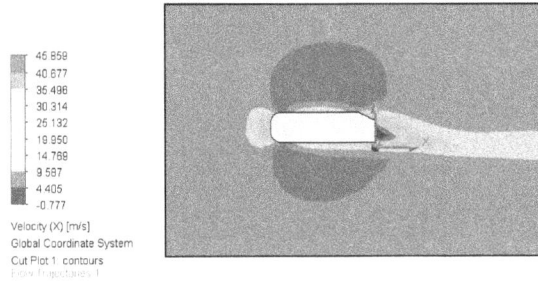

Figure 8-36   *The Velocity (X) cut plot*

Creating the Flow Trajectories

Next, you need to invoke the **Cut Plot** tool and observe the velocity plot.

1. Choose the **Flow Trajectories** tool from the **Insert** drop-down in the **Flow Simulation CommandManager**; the **Flow Trajectories PropertyManager** is displayed, as shown in Figure 8-37.

2. Select the Front Plane from the drawing area, the name of plane is displayed in the **Planes Faces**, **Sketches**, **Edges** and **Curves** selection box in the **Starting Points** rollout. Ensure that the **Pattern** button is chosen in the **Starting Points** rollout.

3. Select the **Lines with Arrows** option from the **Draw Trajectory As** drop-down list from the **Appearance** rollout. Ensure that the **Static Trajectories** button is chosen in the **Appearance** rollout.

4. Select the **Velocity (X)** option from the **Color by** drop-down list in the **Appearance** rollout.

5. Choose the **OK** button to close the PropertyManager; the trajectories are displayed in the drawing area. To view the trajectories, hide the model. You will notice the velocity distribution in the model, refer to Figure 8-38.

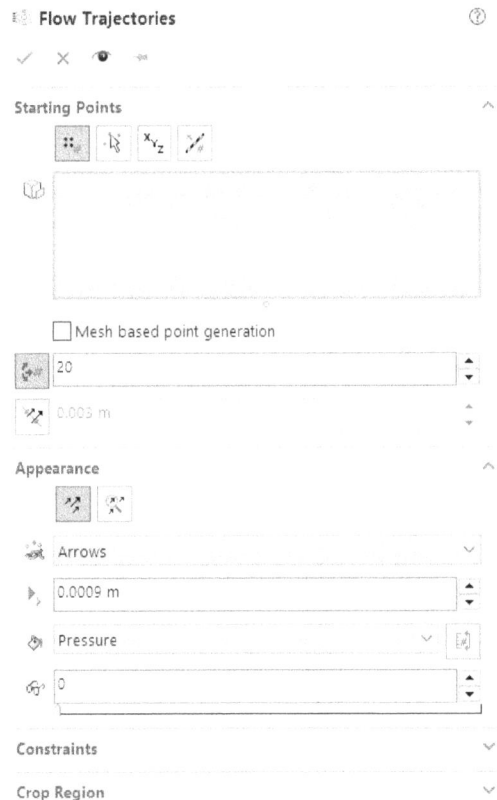

Figure 8-37   *The Flow Trajectories PropertyManager*

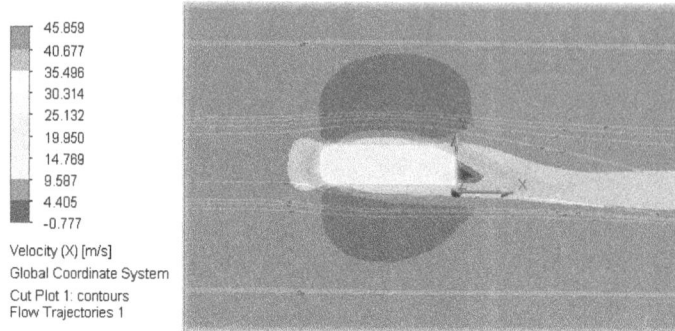

Figure 8-38 *The Velocity(X) flow trajectories*

Creating the Goal Plot

Next, you need to invoke the **Goal Plot** tool and observe the goal values.

1. Choose the **Goal Plot** tool from the **Insert** drop-down in the **Flow Simulation CommandManager**; the **Goal Plot PropertyManager** is displayed, as shown in Figure 8-39.

2. Select all check boxes from the **Goals to Plot** list in the **Goals** rollout.

3. Choose the **Show** button from the **Options** rollout; the **Goal Plot 1** tab is added to the bottom of the drawing area. Here, you will note down the force values.

4. Choose the **OK** button to close the PropertyManager; the Force versus Iterations graph is displayed in the drawing area. You will notice how the force value changes with respect to the iteration, refer to Figure 8-40.

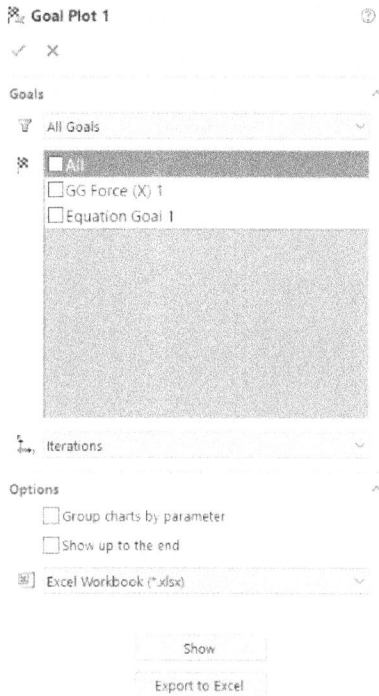

Figure 8-39   *The **Goal Plot** PropertyManager*

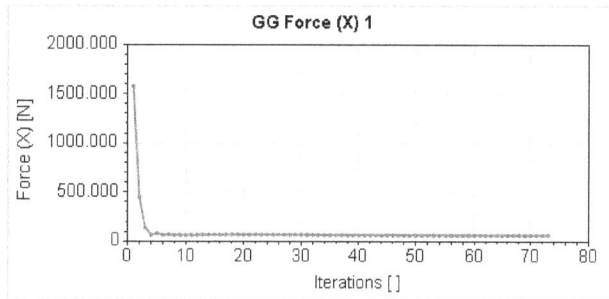

Figure 8-40   *The **Force(X)** goal plot*

Saving the Model

1. Save the part document with the name *c08_tut03* at the following location: *\SOLIDWORKS_ Flow\resources\c08*.

2. Choose **File > Close** from the SOLIDWORKS menus to close the document.

Tutorial 4

In this tutorial, you will open the model (*c07_tut04*) created in Tutorial 4 of Chapter 7. You can also download this file from *www.cadcim.com* by using the following path:

Textbooks > CAE Simulation > Dassault Systemes > SOLIDWORKS Flow Simulation > Flow Simulation Using SOLIDWORKS 2023 > Input Files > C08_SWFS_inp

You will then visualize the results of the simulation. The model is shown in Figure 8-41.

(Expected time: 20 min)

The following steps are required to complete this tutorial:

a. Open Tutorial 4 of Chapter 7.
b. Save this tutorial in the *c08* folder with a new name.

c. Add cut plots to the model.
d. Add flow trajectories to the model.
e. Save the file.

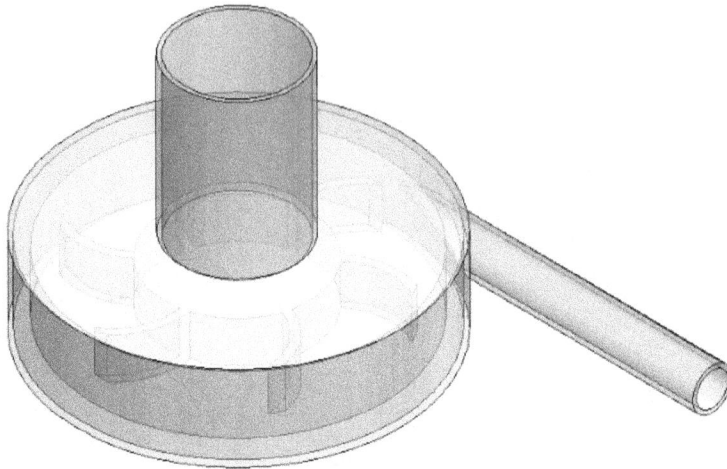

Figure 8-41 *Model for Tutorial 4*

Opening Tutorial 4 of Chapter 7

As the required document is saved in the *c07* folder, you need to select this folder and then open the *c07_tut04.sldprt* document.

1. Start SOLIDWORKS by double-clicking on its shortcut icon on the desktop of your computer.

2. Choose the **Open** button from the Menu Bar to display the **Open** dialog box, refer to Figure 8-42.

3. Browse to the SOLIDWORKS folder and select the **c07** folder.

4. Select the **c07_tut04.sldprt** document and then choose the **Open** button.

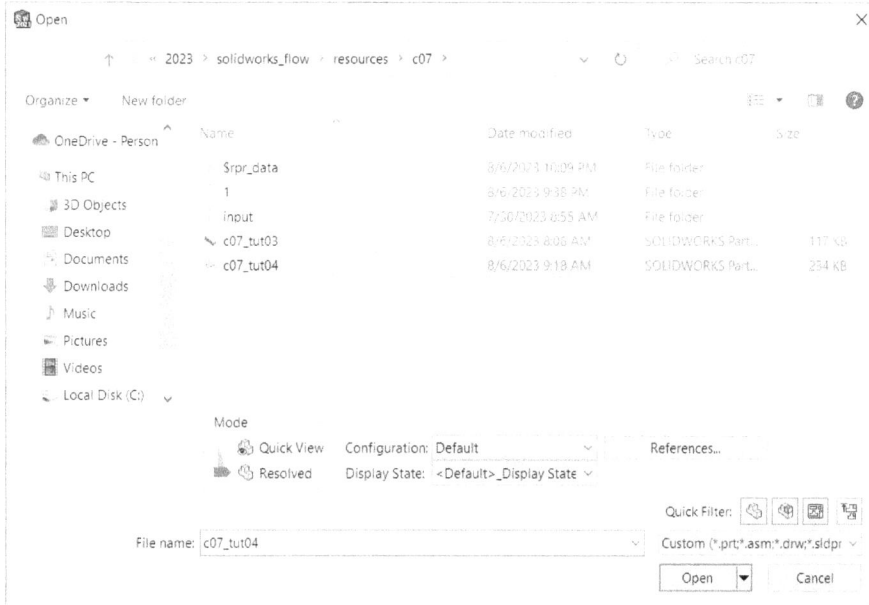

*Figure 8-42   The **Open** dialog box*

Saving the Document

When you open a document from another chapter, it is recommended that you first save the opened document with a new name in the folder of the current chapter to avoid the original document from getting modified.

1. Choose the **Save As** button from the **Save** flyout in the Menu Bar; the **Save As** dialog box is displayed, as shown in Figure 8-43.

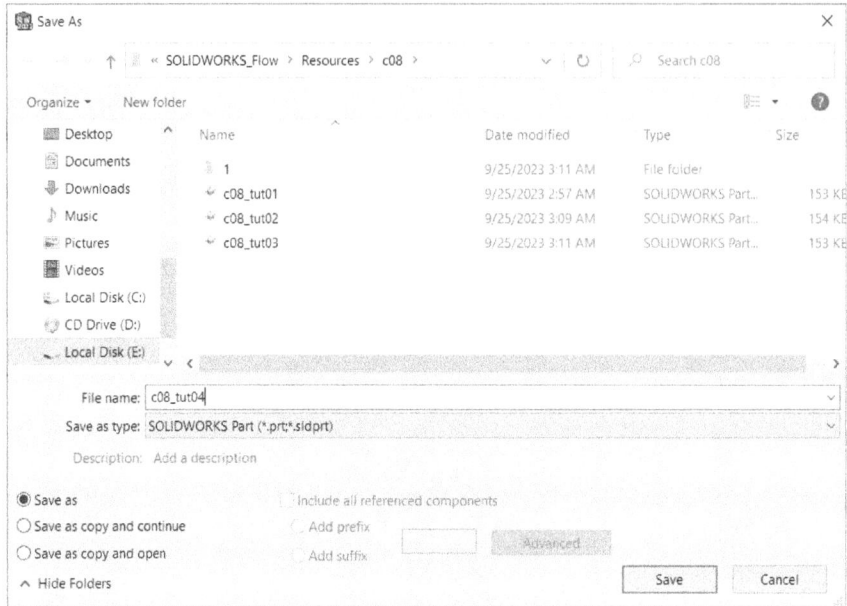

*Figure 8-43 The **Save As** dialog box*

2. Browse to the **SOLIDWORKS_Flow > Resources** folder and make the **c08** folder as the current folder by double-clicking on it.

3. Enter **c08_tut04** as the new name of the document in the **File name** edit box and then choose the **Save** button to save the document. The document is saved with the new name and gets opened in the drawing area.

Running the Calculation

Next, you need to invoke the **Run** tool and start the calculation.

1. Choose the **Run** tool from the **Flow Simulation CommandManager**; the **Run** dialog box is displayed, as shown in Figure 8-44.

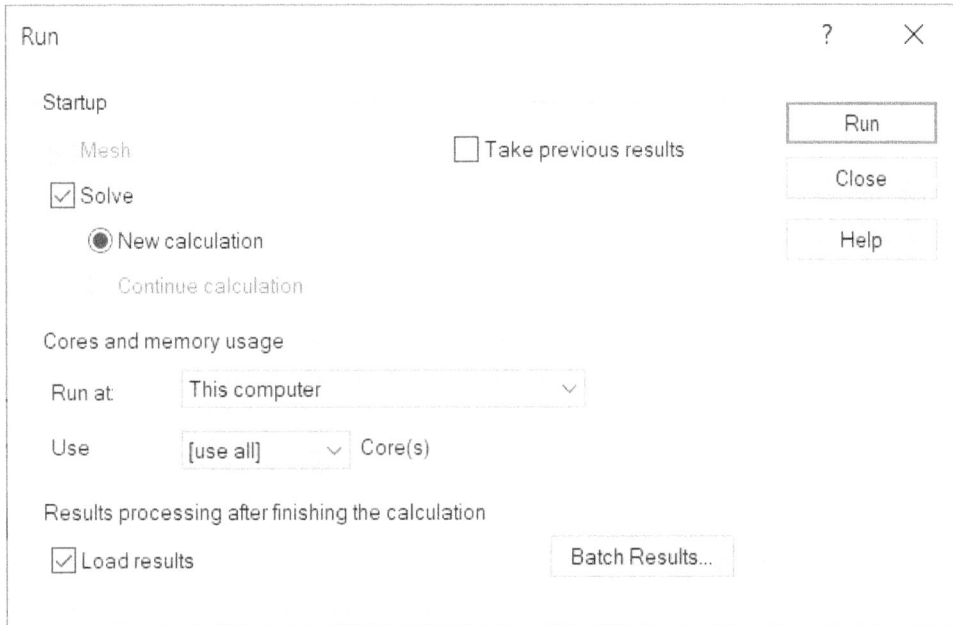

Figure 8-44   *The **Run** dialog box*

2. Select the **Mesh** check box and also ensure that the **Solve** check box is selected by default. Choose the **Run** button; first the **Geometry Preparation**, then **Mesh Generation** and then the **Solver** window are displayed. Note that the window names change as the respective operation finishes. Finally, the **Solver** dialog box is displayed and the "**Solver is finished**" message box is displayed.

3. Now, choose the **Close** button to close the dialog box.

4. Right-click on the **Results** node from the **Flow Simulation Analysis Tree** and select the **Load** option to load the results.

Creating the Cut Plot

Next, you need to invoke the **Cut Plot** tool and observe the velocity plot.

1. Choose the **Cut Plot** tool from the **Insert** drop-down in the **Flow Simulation CommandManager**; the **Cut Plot PropertyManager** is displayed, as shown in Figure 8-45.

2. Select the Top Plane from the drawing area which is displayed in the **Section Plane**, **Planar Face or Curve** selection box in the **Selection** rollout.

3. Choose the **Contours** button from the **Display** rollout if it is not chosen.

4. Select the **Pressure** parameter from the **Parameter** drop-down list in the **Contours** rollout.

5.   Choose the **OK** button to close the PropertyManager; the Pressure cut plot is displayed in the drawing area, refer to Figure 8-46. Note that, to view the cut plot, hide the model from the drawing area. Here, you will notice that the maximum pressure is exerted around the outlet.

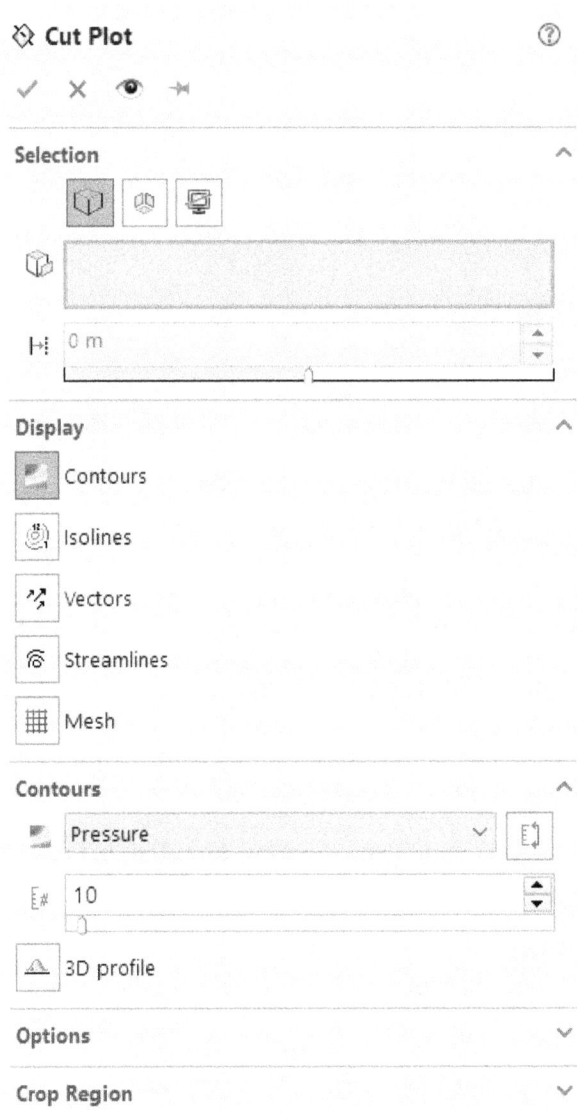

Figure 8-45   The Cut Plot PropertyManager

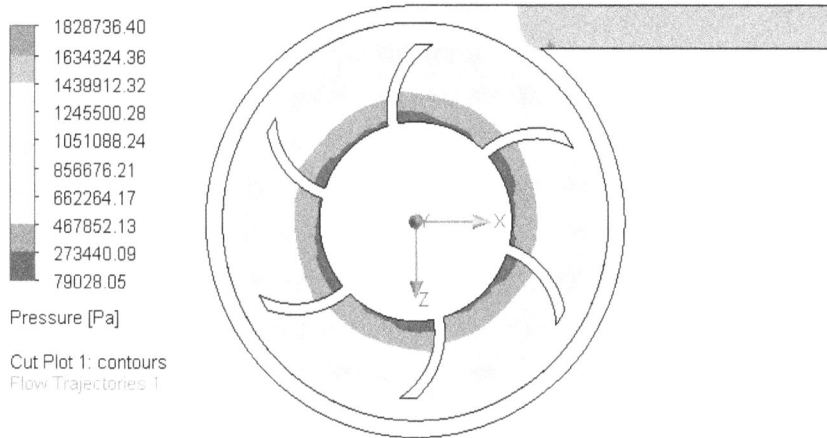

1828736.40
1634324.36
1439912.32
1245500.28
1051088.24
856676.21
662264.17
467852.13
273440.09
79028.05

Pressure [Pa]

Cut Plot 1: contours
Flow Trajectories 1

Figure 8-46   *The Pressure cut plot*

Creating the Flow Trajectories

Next, you need to invoke the **Cut Plot** tool and observe the velocity plot.

1. Choose the **Flow Trajectories** tool from the **Insert** drop-down in the **Flow Simulation CommandManager**; the **Flow Trajectories PropertyManager** is displayed, as shown in Figure 8-47.

2. Select the Inlet and Outlet faces from the drawing area which is displayed in the **Planes Faces, Sketches, Edges and Curves** selection box in the **Starting Points** rollout. Ensure that the **Pattern** button is chosen in the **Starting Points** rollout.

3. Select the **Arrows** option from the **Draw Trajectory As** drop-down list from the **Appearance** rollout. Ensure that the **Static Trajectories** button is chosen in the **Appearance** rollout.

4. Select the **Pressure** option from the **Color by** drop-down list in the **appearance** rollout.

5. Choose the **OK** button to close the PropertyManager; the trajectories are displayed in the drawing area. To view the trajectories, hide the model. You will notice the pressure distribution in the model, refer to Figure 8-48.

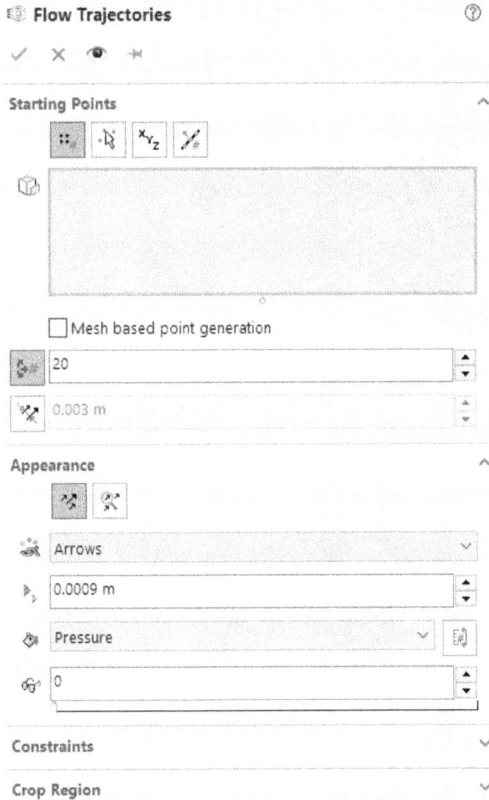

Figure 8-47   The Flow Trajectories
PropertyManager

Figure 8-48   The Pressure flow trajectories

Saving the Model

1. Save the part document with the name *c08_tut04* at the following location: *\SOLIDWORKS_Flow\resources\c08*.

2. Choose **File > Close** from the SOLIDWORKS menus to close the document.

Self-Evaluation Test

Answer the following questions and then compare them to those given at the end of this chapter:

1. The _____ tool is used to start calculation for the current project.

2. The _____ button is used to create cut plot parallel to the planar face.

3. The _____ check box is used to calculate the project.

4. The _____ button is used to evenly distribute the trajectory starting points over the selected planes, faces, sketches, edges and curves.

5. The _____ button allows you to specify a set of starting points by creating a table with points coordinates.

Review Questions

Answer the following questions:

1. Which of the following check boxes should be selected if you want to create a new computational mesh for a project that is already calculated?

    (a) **Computational Mesh**   (b) **Mesh**
    (c) **Global Mesh**   (d) None of these

2. Which of the following buttons is used to create a plane parallel to one of the default planes?

    (a) **XYZ Planes**   (b) **Normal to Screen**
    (c) **Reference**   (d) None of these

3. The _____ button is used to allow you to select starting points on a plane or planar face in the graphics area.

4. The _____ button is used to evenly distribute the trajectory starting points over a line, a rectangle, or a sphere.

5. The **Dynamic Trajectories** button is used to display trajectories as a static image. (T/F)

EXERCISES

Exercise 1

In this exercise, you will open the model (*c07_exr01*) created in Exercise 1 of Chapter 7. You will then apply goals to the model. You can also download this file from *www.cadcim.com* by using the following path:

Textbooks > CAE Simulation > Dassault Systemes > SOLIDWORKS Flow Simulation > Flow Simulation Using SOLIDWORKS 2023 > Input Files > C08_SWFS_inp

The model is shown in Figure 8-49.

(Expected time: 20 min)

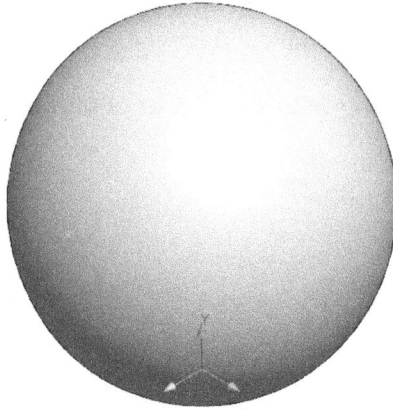

Figure 8-49 Solid model for Exercise 1

Exercise 2

In this exercise, you will open the model (*c07_exr02*) created in Exercise 2 of Chapter 7. You will then apply goals to the model. You can also download this file from *www.cadcim.com* by using the following path:

Textbooks > CAE Simulation > Dassault Systemes > SOLIDWORKS Flow Simulation > Flow Simulation Using SOLIDWORKS 2023 > Input Files > C08_SWFS_inp

The model is shown in Figure 8-50.

(Expected time: 20 min)

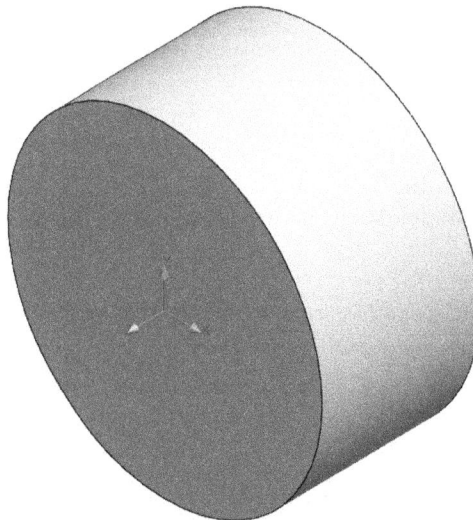

Figure 8-50 Solid model for Exercise 2

Answers to Self-Evaluation Test
1. Run, 2. Reference, 3. Solve, 4. Pattern, 5. Coordinates

Index

Other Publications by CADCIM Technologies

The following is the list of some of the publications by CADCIM Technologies. Please visit *www.cadcim.com* for the complete listing.

AutoCAD Textbooks
- AutoCAD 2024: A Problem-Solving Approach, Basic and Intermediate, 30th Edition
- AutoCAD 2023: A Problem-Solving Approach, Basic and Intermediate, 29th Edition
- AutoCAD 2022: A Problem-Solving Approach, Basic and Intermediate, 28th Edition
- AutoCAD 2021: A Problem-Solving Approach, Basic and Intermediate, 27th Edition
- Advanced AutoCAD 2024: A Problem-Solving Approach (3D and Advanced), 27th Edition

Autodesk Inventor Textbooks
- Autodesk Inventor Professional 2024 for Designers, 24th Edition
- Autodesk Inventor Professional 2023 for Designers, 23rd Edition
- Autodesk Inventor Professional 2022 for Designers, 22nd Edition
- Autodesk Inventor Professional 2021 for Designers, 21st Edition

AutoCAD MEP Textbooks
- AutoCAD MEP 2023 for Designers, 7th Edition
- AutoCAD MEP 2022 for Designers, 6th Edition
- AutoCAD MEP 2020 for Designers, 5th Edition

AutoCAD Plant 3D Textbooks
- AutoCAD Plant 3D 2024 for Designers, 8th Edition
- AutoCAD Plant 3D 2023 for Designers, 7th Edition
- AutoCAD Plant 3D 2021 for Designers, 6th Edition

Autodesk Fusion 360 Textbook
- Autodesk Fusion 360: A Tutorial Approach, 4th Edition
- Autodesk Fusion 360: A Tutorial Approach, 3rd Edition

Solid Edge Textbooks
- Solid Edge 2023 for Designers, 20th Edition
- Solid Edge 2022 for Designers, 19th Edition
- Solid Edge 2021 for Designers, 18th Edition

NX Textbooks
- Siemens NX 2021 for Designers, 14th Edition
- Siemens NX 2020 for Designers, 13th Edition
- Siemens NX 2019 for Designers, 17th Edition

NX Mold Textbook
- Mold Design Using NX 11.0: A Tutorial Approach

NX Nastran Textbook
• NX Nastran 9.0 for Designers

SOLIDWORKS Textbooks
• SOLIDWORKS 2023 for Designers, 21st Edition
• Advanced SOLIDWORKS 2022 for Designers, 20th Edition
• SOLIDWORKS 2022: A Tutorial Approach, 6th Edition
• Learning SOLIDWORKS 2022: A Project Based Approach

SOLIDWORKS Simulation Textbooks
• SOLIDWORKS Simulation 2022: A Tutorial Approach
• SOLIDWORKS Simulation 2018: A Tutorial Approach

CATIA Textbooks
• CATIA V5-6R2022 for Designers, 20th Edition
• CATIA V5-6R2021 for Designers, 19th Edition
• CATIA V5-6R2020 for Designers, 18th Edition

Creo Parametric Textbooks
• Creo Parametric 9.0 for Designers, 9th Edition
• Creo Parametric 8.0 for Designers, 8th Edition

ANSYS Textbooks
• ANSYS Workbench 2023 R2: A Tutorial Approach
• ANSYS Workbench 2022 R1: A Tutorial Approach
• ANSYS Workbench 2021 R1: A Tutorial Approach

Creo Direct Textbook
• Creo Direct 2.0 and Beyond for Designers

Autodesk Alias Textbooks
• Learning Autodesk Alias Design 2016, 5th Edition
• Learning Autodesk Alias Design 2015, 4th Edition

AutoCAD LT Textbooks
• AutoCAD LT 2024 for Designers, 16th Edition
• AutoCAD LT 2023 for Designers, 15th Edition

EdgeCAM Textbooks
• EdgeCAM 11.0 for Manufacturers
• EdgeCAM 10.0 for Manufacturers

Autodesk Revit MEP Textbooks
• Exploring Autodesk Revit 2023 for MEP, 9th Edition
• Exploring Autodesk Revit 2022 for MEP, 8th Edition

AutoCAD Civil 3D Textbooks
* Exploring AutoCAD Civil 3D 2023, 12th Edition
* Exploring AutoCAD Civil 3D 2022, 11th Edition
* Exploring AutoCAD Civil 3D 2021, 10th Edition

AutoCAD Map 3D Textbooks
* Exploring AutoCAD Map 3D 2023, 10th Edition
* Exploring AutoCAD Map 3D 2022, 9th Edition
* Exploring AutoCAD Map 3D 2018, 8th Edition

RISA-3D Textbook
* Exploring RISA-3D 14.0

Autodesk Navisworks Textbooks
* Exploring Autodesk Navisworks 2023, 10th Edition
* Exploring Autodesk Navisworks 2022, 9th Edition
* Exploring Autodesk Navisworks 2021, 8th Edition

AutoCAD Raster Design Textbooks
* Exploring AutoCAD Raster Design 2017
* Exploring AutoCAD Raster Design 2016

Bentley STAAD.Pro Textbooks
* Exploring Bentley STAAD.Pro CONNECT Edition, 5th Edition
* Exploring Bentley STAAD.Pro CONNECT Edition, 4th Edition
* Exploring Bentley STAAD.Pro (CONNECT) Edition

Autodesk 3ds Max Design Textbooks
* Autodesk 3ds Max Design 2015: A Tutorial Approach, 15th Edition
* Autodesk 3ds Max Design 2014: A Tutorial Approach

Autodesk 3ds Max Textbooks
* Autodesk 3ds Max 2023 for Beginners: A Tutorial Approach, 23rd Edition
* Autodesk 3ds Max 2022 for Beginners: A Tutorial Approach, 22nd Edition
* Autodesk 3ds Max 2021: A Comprehensive Guide, 21st Edition
* Autodesk 3ds Max 2020: A Comprehensive Guide, 20th Edition

Autodesk Maya Textbooks
* Autodesk Maya 2024: A Comprehensive Guide, 15th Edition
* Autodesk Maya 2023: A Comprehensive Guide, 14th Edition
* Autodesk Maya 2022: A Comprehensive Guide, 13th Edition

Pixologic ZBrush Textbooks
* MAXON ZBrush 2023: A Comprehensive Guide, 9th Edition
* Pixologic ZBrush 2022: A Comprehensive Guide, 8th Edition
* Pixologic ZBrush 2021: A Comprehensive Guide, 7th Edition

Fusion Textbooks
- Blackmagic Design Fusion 7 Studio: A Tutorial Approach, 3rd Edition
- The eyeon Fusion 6.3: A Tutorial Approach

Flash Textbooks
- Adobe Flash Professional CC 2015: A Tutorial Approach, 3rd Edition
- Adobe Flash Professional CC: A Tutorial Approach

Computer Programming Textbooks
- Introducing PHP 7/MySQL
- Introduction to C++ programming, 2nd Edition
- Learning Oracle 12c - A PL/SQL Approach
- Learning ASP.NET AJAX
- Introduction to Java Programming, 2nd Edition
- Learning Visual Basic.NET 2008

MAXON CINEMA 4D Textbooks
- MAXON CINEMA 4D R25 : A Tutorial Approach, 9th Edition
- MAXON CINEMA 4D S24: A Tutorial Approach, 8th Edition
- MAXON CINEMA 4D R20 Studio: A Tutorial Approach, 7th Edition

Oracle Primavera Textbooks
- Exploring Oracle Primavera P6 Professional 18, 3rd Edition
- Exploring Oracle Primavera P6 v8.4

AutoCAD Textbooks Authored by Prof. Sham Tickoo and Published by Autodesk Press
- AutoCAD: A Problem-Solving Approach: 2013 and Beyond
- AutoCAD 2012: A Problem-Solving Approach
- AutoCAD 2011: A Problem-Solving Approach
- AutoCAD 2010: A Problem-Solving Approach
- Customizing AutoCAD 2020

Coming Soon from CADCIM Technologies
- ANSYS 2023R2: A Comprehensive Guide

Online Training Program Offered by CADCIM Technologies
CADCIM Technologies provides effective and affordable virtual online training on animation, architecture, and GIS softwares, computer programming languages, and Computer Aided Design, Manufacturing, and Engineering (CAD/CAM/CAE) software packages. The training will be delivered 'live' via Internet at any time, any place, and at any pace to individuals, students of colleges, universities, and CAD/CAM/CAE training centers. For more information, please visit the following link: ***https://www.cadcim.com***.

www.ingramcontent.com/pod-product-compliance
Lightning Source LLC
Chambersburg PA
CBHW081807200326
41597CB00023B/4182